# MORE *than just a man*

## A Dramatic Easter Musical

*Created by* Chris Machen and Kim Messer

*Arranged and Orchestrated by* Richard Kingsmore

lillenas.com

# CONTENTS

# Cast

| | |
|---|---|
| YOUNG JESUS | Jesus at the time of his youthful visit to the temple to teach |
| YOUNG JOSEPH | Joseph of Arimathea in his early 30s; nicely dressed in civilian garb; this role can be combined with the older Joseph if needed |
| YOUNG CAIAPHAS | Caiaphas in his early 30s; also nicely dressed in civilian garb; this role can be combined with the older Caiaphas if needed |
| MARY | mother of Jesus; appears both when Jesus is a young boy at the temple and at His crucifixion |
| JOSEPH *(non-speaking)* | Jesus' earthly father; finds Young Jesus at the temple |
| JOSEPH (OF ARIMATHEA) | council member and best friend to Caiaphas; in his early 50s; wears a council robe |
| CAIAPHAS | the High Priest; leader of the council; in his early 50s; wears a distinguishing council robe |
| CALEB | child in Jerusalem |
| LEAH | child in Jerusalem |
| SHANNON | child in Jerusalem |
| LIAM | child in Jerusalem |
| NICODEMUS | council member, and learned man in his late 50s; wears council robes |
| NAOMI | crowd member |
| NILA | crowd member |
| ZEPHANIAH | crowd member |
| ELIAKIM | council member; wears council robes |
| JONATHAN | council member; wears council robes |
| PEKAH | council member; wears council robes |
| JORAM | council member; wears council robes |
| PETER | one of Jesus' disciples |
| MARY MAGDALENE | witness to the resurrection |
| JOANNA | accompanies Mary Magdalene |
| EXTRAS | Townspeople, Roman Soldiers, Disciples, Council Members, Priests |

# Scene 1

*Luke 2:49-50*

*(During* CALL TO WORSHIP, *the choir is on stage. It's a typical day in the streets– merchants outside the temple, women talking to each other, children playing and laughing, etc. A* YOUNG JESUS *is in the temple speaking with the* PRIESTS, *who are amazed at what He has to say.* YOUNG CAIAPHAS *and* YOUNG JOSEPH *are present and see the young priest in action.)*

## Call to Worship

Words and Music by
CHRIS and DIANE MACHEN
*Arranged by Richard Kingsmore*

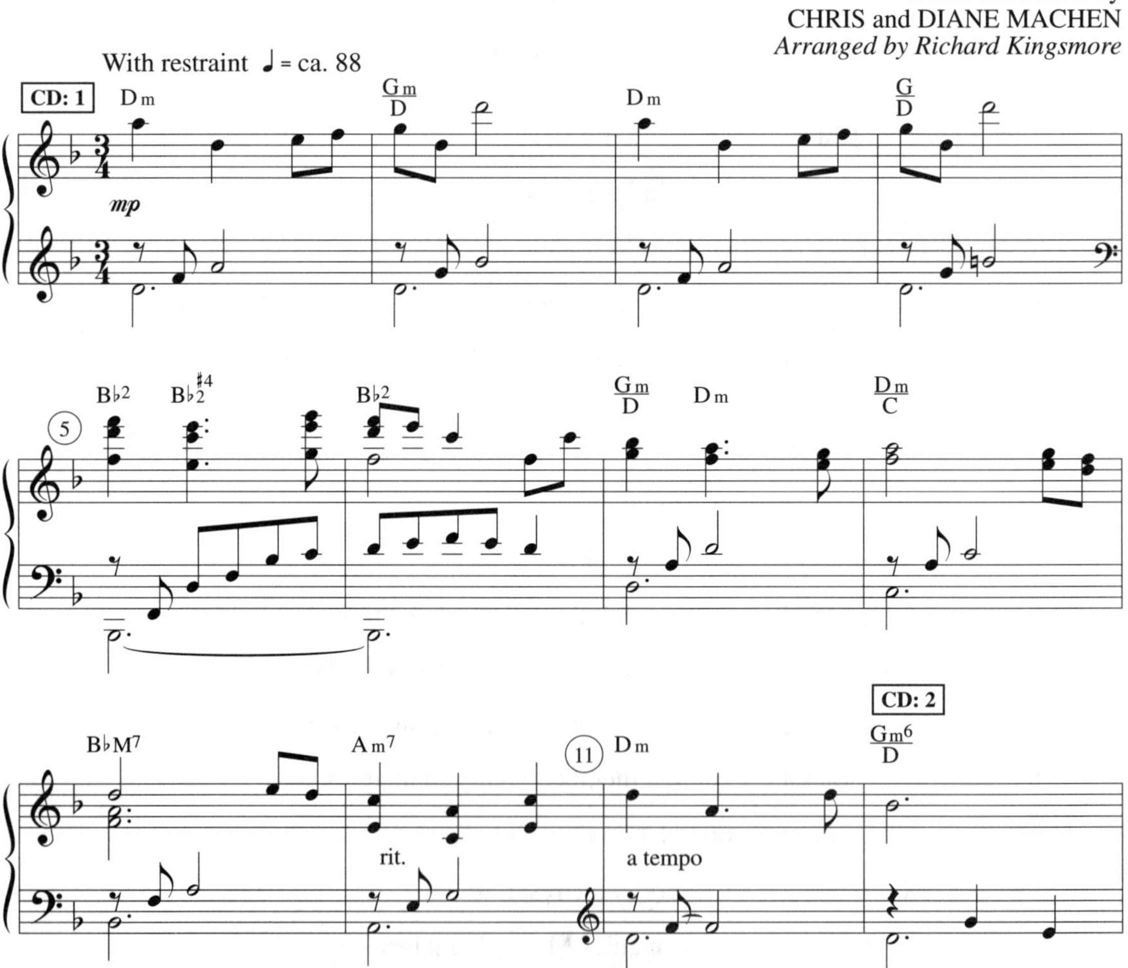

15 *Child solo*
*mp*

Come to the house of the Lord, to the house of the Lord,

20

Bring - ing your praise to the One we a - dore;

24

For the Lord, He is God, Let us rise and ap - plaud;

28

Let us wor - ship His

Dm | Gm6/D | Dm | Gm6/D
Dm | G/D | Dm | C
A | A7/C♯ | Dm | Gm7/D
Dm | B♭ | Dm/A | Gm

rit.
31
Faster ♩. = ca. 106
name, He is here, stand in awe.
Am7
A7
N.C.
rit.
Faster
35
Dm
f
C
B♭
B♭
C
39
Dm
C
CD: 3
B♭
C

1st time: Choir
2nd time: Ladies
43
mf
1.Come to the house of the Lord, to the house of the
2.Come to the house of the Lord, to the house of the
Dm
Gm/D
G/D
Lord, Bring - ing your praise to the One we a - dore;
Lord; Lift high your praise to the King ev - er - more.
C
A
A7/C♯
47
Both times: Choir
For the Lord, He is God, Let us rise and ap -
In His love stand se - cure, for His mer - cy en -
Gm7/D
B♭

plaud; Let us wor - ship His name, He is here, stand in
dures; Let us wor - ship and sing from a heart that is
Dm/A
Gm
A
A7
51
cresc.
CD: 4 / 6 1st / 2nd time
awe. Come bring your of - fer - ing.
pure. On - ly Je - ho - vah reigns.
B♭
C6
C
Come and ex - alt and sing.
Come now ex - alt His name.
Re - joice!
f
Gm
C6
C

55

Lift up your voice. Sing to the

55 D D/F♯ Dsus/F♯ D/F♯ Gm B♭ F/B♭ Gm/B♭

*f*

Lord here in His sanc - tu - ar - y.

C C4/2 C A7♭9 A7

59

Rise! Lift up your eyes, hon - or His

59 D D/F♯ Dsus/F♯ D/F♯ Gm B♭ F/B♭ Gm/B♭

name, Bring praise and glo - ry to the
C C 4/2 C A 7 b9 A 7
63
Lord of all.
Gm/Bb Bb
Hear the call to
C 6 C

1

wor - ship!

1 C D m C B♭ C

CD: 5

(to pg. 7, meas. 43)

D m C B♭ C

(to pg. 7, meas. 43)

2 CD: 7

wor - ship!

2 C D

73
En - ter His gates with thanks - giv - ing.
A
A2
A
Bm
En - ter His courts with praise.
F♯m
G
G♯4 2
G
77
CD: 8
Wor - ship the ho - ly mag - nif - i - cent
Em7
D/F♯

cresc.

*ff*

God. Re - joice!

cresc.

*ff*

A sus A B sus B

cresc.

81

Lift up your voice. Sing to the

81

E $\frac{E}{G\sharp}$ $\frac{E\ sus}{G\sharp}$ $\frac{E}{G\sharp}$ A m C $\frac{G}{C}$ $\frac{A\,m}{C}$

*ff*

Lord here in His sanc - tu - ar - y.

D D $^{4}_{2}$ D B $^{\flat 9}_{7}$ B $^{7}$

85
Rise! Lift up your eyes, hon - or His
E
E/G♯
E sus/G♯
E/G♯
Am
C
G/C
Am/C
name, Bring praise and glo - ry to the
D
D 4/2
D
B 7 ♭9
B 7
89
Lord of all.
Am/C
C

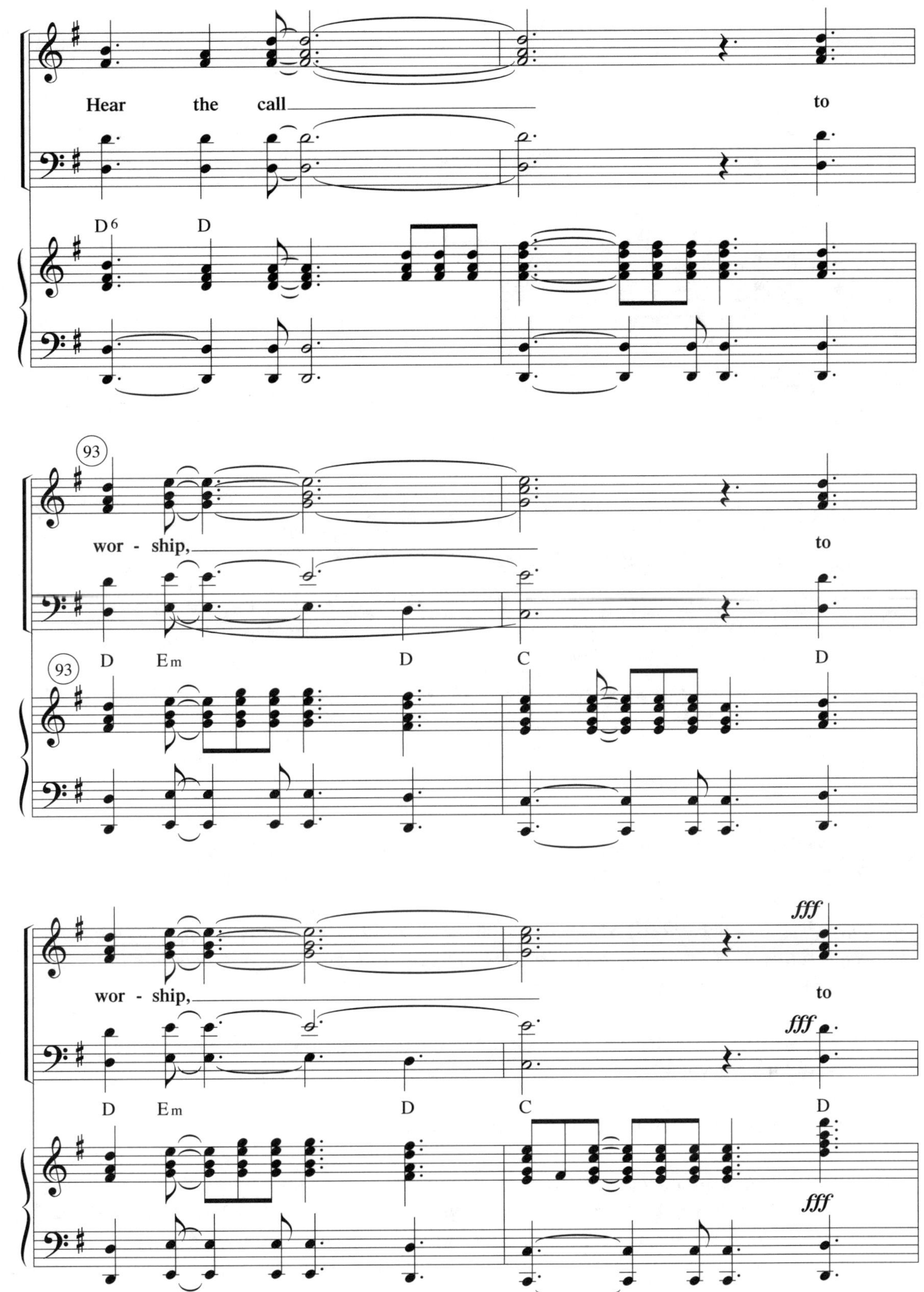
Hear the call
to
D6
D
93
wor - ship,
to
D
Em
D
C
D
wor - ship,
fff
to
fff
D
Em
D
C
D
fff

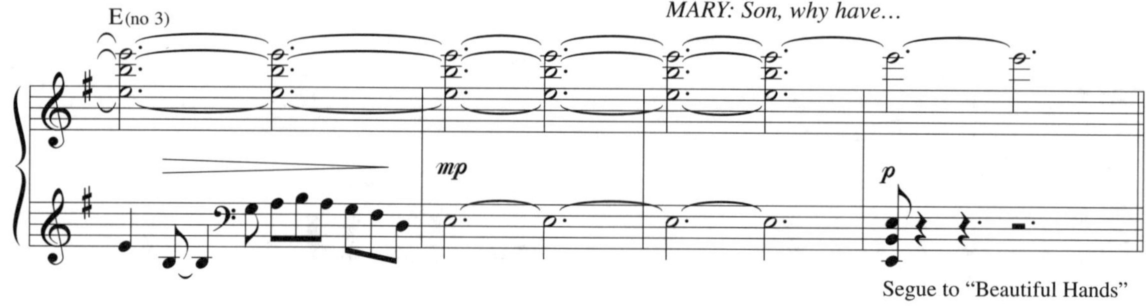


*(At approximately measure 91,* MARY, *Jesus' mother, and* JOSEPH, *Jesus' earthly father, enter the scene frantically looking for their boy. They ask around. Someone has seen him in the temple.* JOSEPH *finds* YOUNG JESUS *and leads him outside to* MARY.*)*

MARY: Son, why have you treated us like this? Your father and I have been anxiously searching for you.

JESUS: Why were you searching for me? Didn't you know I had to be in my Father's house?

*(*MARY *is relieved and disturbed that they have found* JESUS. *She takes His hands in hers and sings* BEAUTIFUL HANDS (PART 1). JOSEPH *hugs* YOUNG JESUS, *then* MARY, JOSEPH *and* YOUNG JESUS *exit as the song concludes, and those in the temple begin to disperse and go their separate ways.* YOUNG CAIAPHAS *and* YOUNG JOSEPH *talk as they leave the temple, discussing what their impressions are of this young Jesus.)*

# Beautiful Hands
(Part 1)

Words and Music by
CHRIS MACHEN
*Arranged by Richard Kingsmore*

11
CD: 10
rit.
heart can't un - der - stand How beau - ti - ful these pre - cious
A 2(no 3)
D m9
E sus
rit.
a tempo
16
mf
hands. Beau - ti - ful hands that touch my
Ladies
mf
Oo
A 2(no 3)
16
A m2
a tempo
mf
face, Beau - ti - ful hands that I em - brace. Beau - ti - ful
A m2
A m
G
F 2
G 2

20
Child, Your Moth- er's here, And I will al - ways hold You,
Ah
Dm Am Dm9 G2
decresc.
p rit.
I will al - ways hold You
decresc.
Hold You,
A 4 2 Am Dm9 G2
rit.
26 a tempo rit.
near.
A 2(no 3) Em11/A FM9/A Em11/A A 2(no 3)
a tempo rit.

YOUNG CAIAPHAS *(boisterously)*: Can you believe that child?

YOUNG JOSEPH *(laughing)*: He was something. Definitely knew his stuff.

YOUNG CAIAPHAS: Yes, but c'mon! Someone must have put him up to it. He couldn't have known all that on his own.

YOUNG JOSEPH *(thinking)*: I suppose not.

YOUNG CAIAPHAS *(being cocky)*: Learned men such as ourselves–now we know a thing or two. It will be no time before we sit in the Sanhedrin. No time at all!

YOUNG JOSEPH: You are so sure of yourself. But…I don't know. It's a big appointment to be on the council. It's the appointment of a lifetime.

YOUNG CAIAPHAS: Believe me my friend; we will see the day come. And sooner than later!

*(The two shake hands and freeze; lights go out on the scene.)*

# Underscore 1

(Beautiful Hands)

CHRIS MACHEN
*Arranged by Richard Kingsmore*

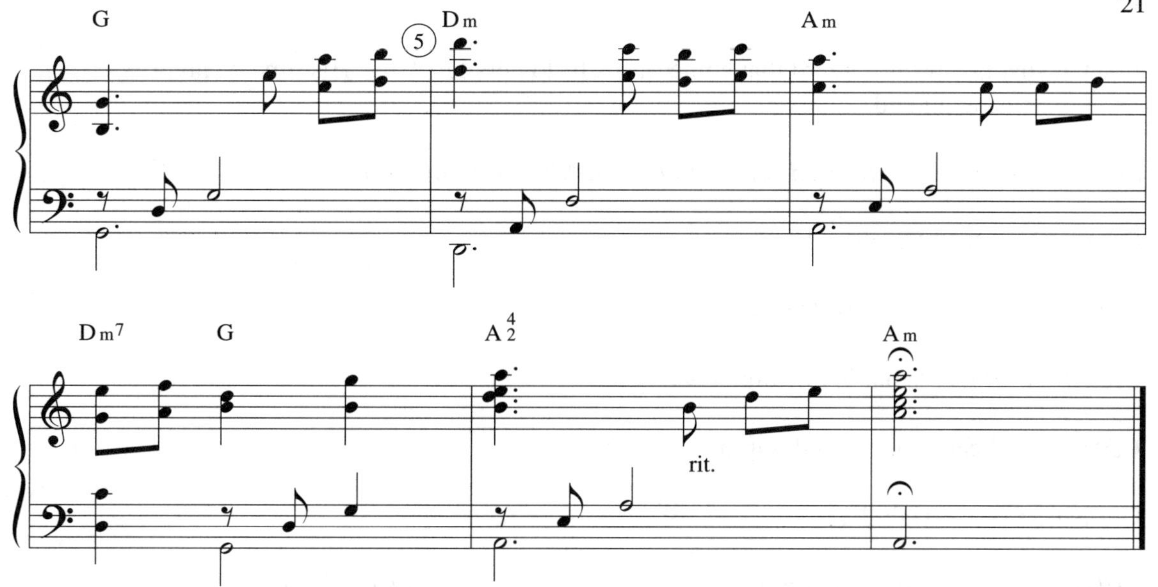


## Scene 2

*John 6 & 7*

*(Scene begins with* JOSEPH *and* CAIAPHAS *in the same exact position as the end of Scene 1. It is 20 years later in Jerusalem.* JOSEPH *and* CAIAPHAS *are now both members of the Sanhedrin.* CAIAPHAS *is the High Priest and* JOSEPH *sits on the council. The characters come out of their "freeze" of shaking hands as the lights come up. They are both dressed in their council robes. Townspeople are bustling about. Everyone is celebrating a feast.)*

CAIAPHAS: My friend! Good to see you.

JOSEPH: And how is the High Priest on this fine day?

CAIAPHAS: This Feast is one of my favorites. I hate to see it end. *(referring to a vendor outside the temple)* Minah always makes the finest treats, and I pretend that she makes them just for me.

JOSEPH: Actually friend, she makes them for me! *(They laugh together. Then* JOSEPH *grows more serious.)* Are you going to the temple to hear Jesus?

CAIAPHAS: Yes, I am. My concerns are growing about this so-called messiah. Yesterday He caused a great commotion. No doubt you heard about it.

JOSEPH: Yes, but we've seen many false prophets come and go. He will most certainly be like the rest.

CAIAPHAS: I don't know about that.

JOSEPH: Just last week I heard that some of his followers left Him. His time of favor is nearing an end.

*(Suddenly a group of children run up to them as they are talking.)*

CALEB: Caiaphas! Caiaphas!

SHANNON: Have you heard? Have you?

CAIAPHAS: What? What is this big news?

LIAM: Jesus healed our friend Sarah.

CAIAPHAS: Is that right?

*(All of the kids respond saying things like: "Yes!" "It was amazing." etc.)*

CAIAPHAS: Did you see it happen?

LEAH: Yes! She can walk now.

CAIAPHAS *(not wanting the kids to see his discomfort)*: Thank you for that helpful information. Run along now. Be careful all of you.

*(They leave to tell others about the miracle. As they exit,* NICODEMUS *enters and approaches* JOSEPH *and* CAIAPHAS.*)*

JOSEPH: Interesting news from the mouths of babes.

NICODEMUS: Hello, friends.

JOSEPH: Hello, Nicodemus.

CAIAPHAS: Tell us you bring news other than what we've heard today.

NICODEMUS: It depends on what you've heard. I've only heard news of Jesus.

CAIAPHAS *(concerned)*: Who is this man? *(music begins)*

JOSEPH: Another so-called messiah like you said.

CAIAPHAS: So why is everyone talking about him?

*(During* WHO IS THIS MAN? PART 1, JESUS *enters the temple area with His* DISCIPLES *and begins teaching. As the crowd becomes aware of His presence, they pay more and more attention to Him.* CAIAPHAS, JOSEPH *and* NICODEMUS *watch intently.)*

# Who Is This Man?

Part 1

Words and Music by
CHRIS MACHEN
*Arranged by Richard Kingsmore*

Driving pulse ♩ = ca. 132

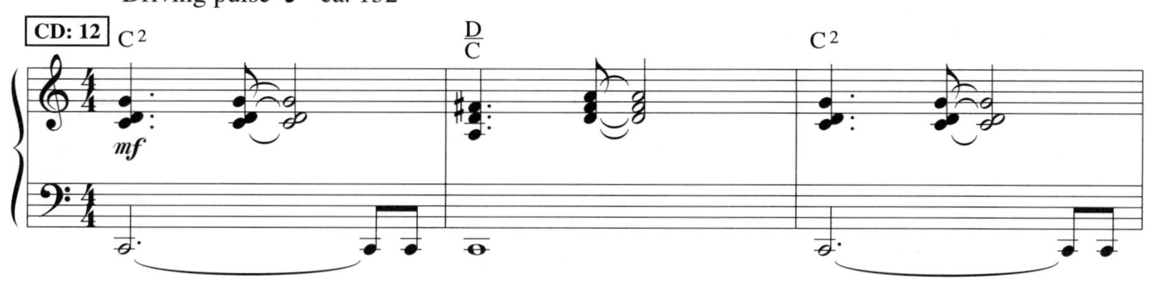

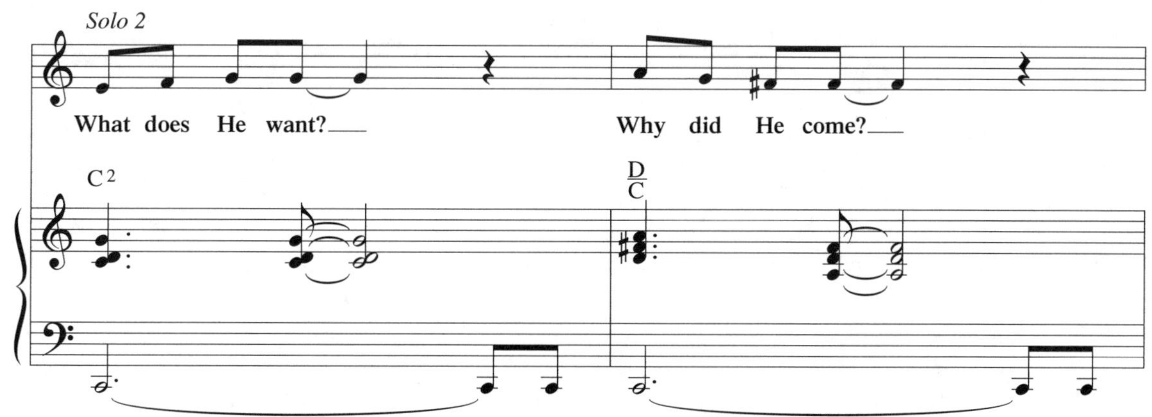

9
Solo 3
Nev - er have we heard such preach - ing. He
Dm7
C/F
G2
speaks with au - thor - i - ty.
13
Solo 4
Not like our scribes or our teach - ers. His
words are of truth and of peace and of pow - er.
CD: 13
G/A
Am

19
Choir
mf
This Je - sus! This
mf
A m9
19
F M7

23
Je - sus! Who is this man?
G
G 2
23
A m

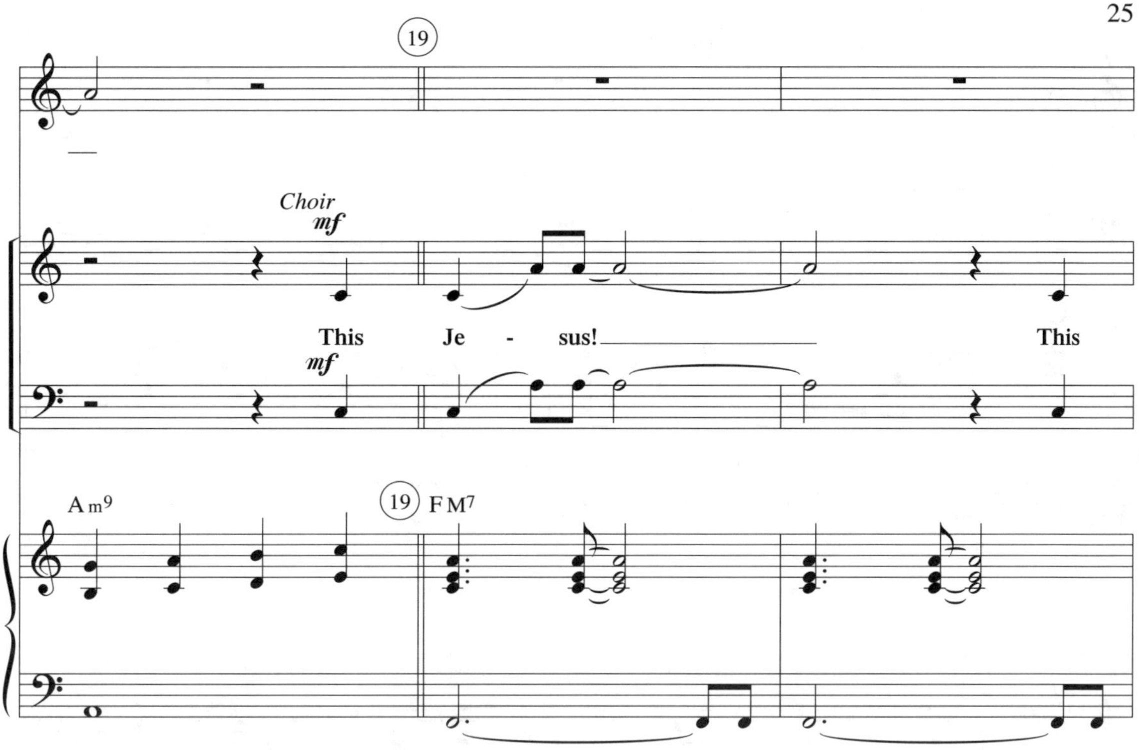

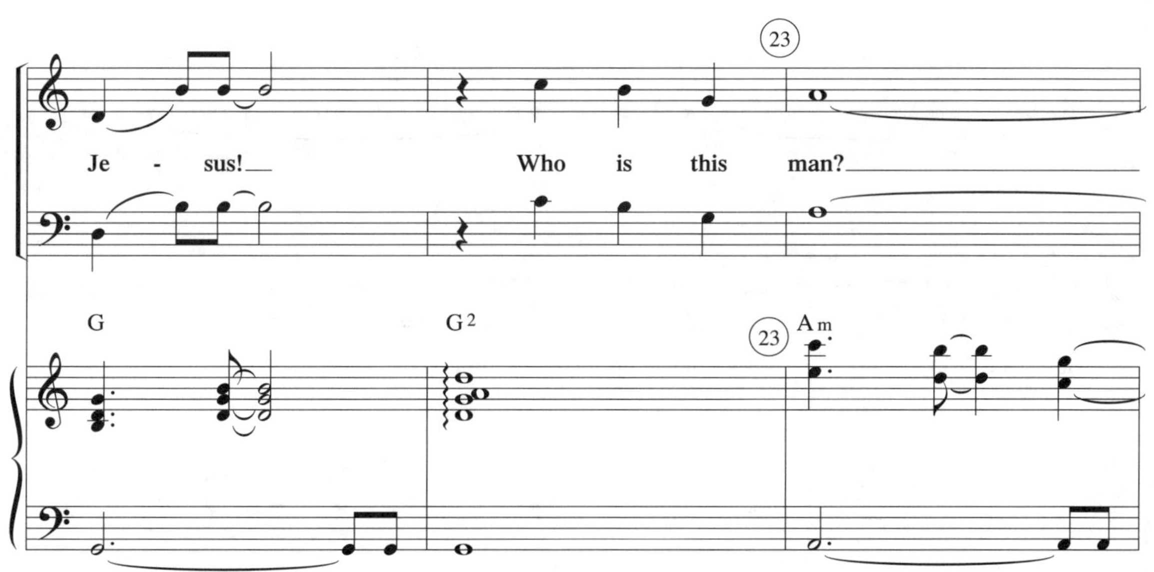

CD: 14
Am
F2
G
27
Solo 5
mf
Who is this man? Look what He's done.
C2
D/C
Solo 6
He can do won - ders like no oth - er one.
C2
D/C

(31) *Solo 7*

Mir - a - cles hap - pen by His hand. The

Dm7 C/F G2

blind are re - ceiv - ing their sight.

Dm7 C/F G2

(35) *Solo 8*

Is He the One, the Mes - si - ah,

Dm7 C/F G2

**CD: 15**

com - ing in pow - er and might for this

Dm7 C/F G2

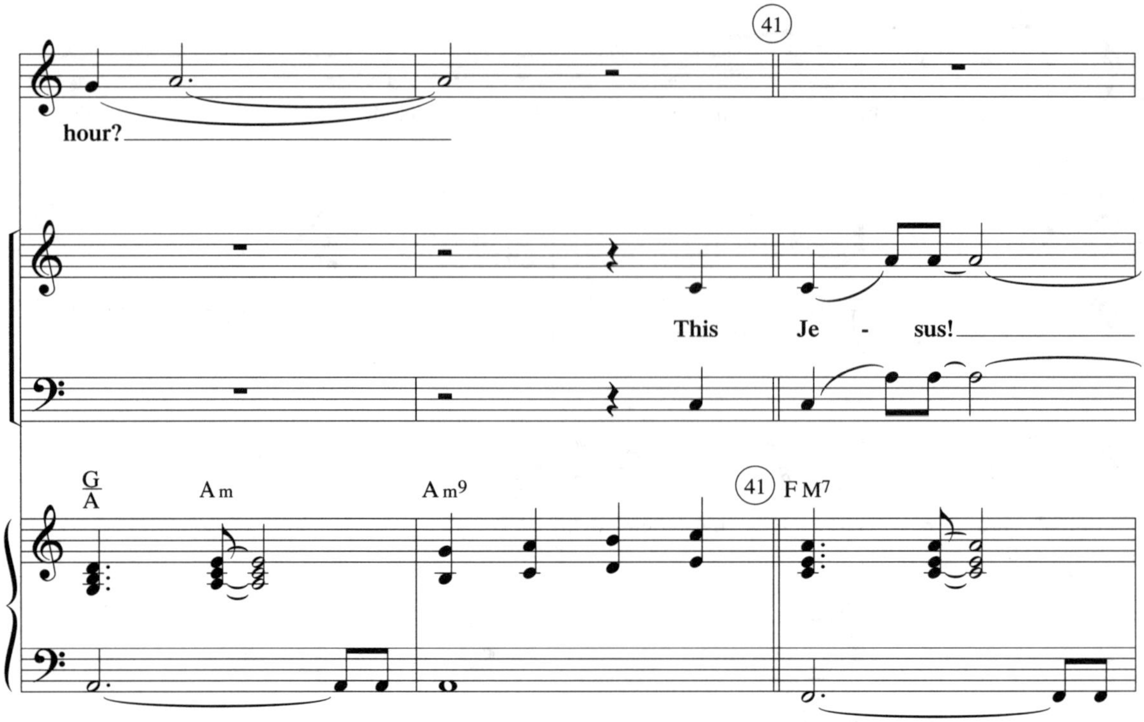

41

hour?

This Je - sus!

G/A Am Am9

41 FM7

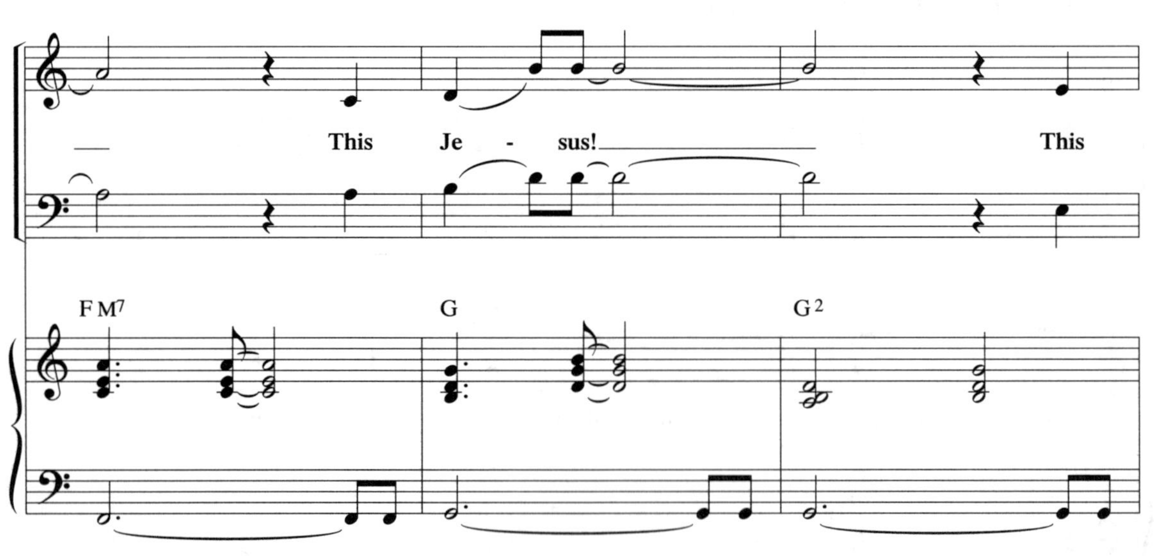

45
Je - sus!
Who
FM7
G2
G sus
49
is this man?
G2
G
G/A
Am
Am9

JESUS *(in a loud voice)*: If anyone is thirsty, let him come to me and drink. Whoever believes in me, as the Scripture has said, streams of water will flow from within him.

*(At this, the crowd cheers and then continues in a celebration of the Feast.* JESUS *exits after a few moments with some following behind. A small group of people near* CAIAPHAS *talks about* JESUS.*)*

NAOMI: Surely this man is the Prophet.

ZEPHANIAH: He is the Christ.

NILA: How can the Christ come from Galilee?

*(They continue their conversation as they exit saying things such as, "I believe Him," "The Messiah comes from Bethlehem, not Galilee," etc. The crowd begins to break up and go their separate ways.)*

CAIAPHAS *(upset)*: You heard what He said. Jesus is definitely a problem.

JOSEPH: The people are divided. I'm telling you, His following is shrinking.

CAIAPHAS: I don't want this kind of trouble. We have a duty to keep the peace. Joseph and Nicodemus, I want you to follow Him.

NICODEMUS: Caiaphas, could you assign someone else from the council to this task?

CAIAPHAS: Not unless you can give me a good reason why you can't do this for me. *(*NICODEMUS *is silent.)* All right then. Jesus will be the topic of discussion at our regular council meeting next week.

*(*NICODEMUS *is visibly uncomfortable with this assignment and distances himself from them just a bit.)*

JOSEPH *(an aside to* CAIAPHAS*)*: Caiaphas, I trust you, but are you sure He's worth the effort?

CAIAPHAS: Yes.

JOSEPH *(looking at* NICODEMUS*)*: Then we'll do this for you. *(pausing and addressing* CAIAPHAS *again)* My friend, do not worry about this. We will handle it.

*(*CAIAPHAS *looks at the two men for a moment then exits.* Joseph *and* NICODEMUS *exit opposite as the lights go to blackout.)*

# Underscore 2

(Nicodemus and Jesus)

RICHARD KINGSMORE
*Arranged by Richard Kingsmore*

# Scene 3

*John 3*

*(It is night now.* JESUS *is sitting and praying alone.* NICODEMUS *enters the scene. He approaches* JESUS *carefully.)*

NICODEMUS: Rabbi, forgive the disturbance. We know you are a teacher who has come from God. No one could do the miracles you do if God were not with him.

JESUS: I tell you the truth, no one can see the kingdom of God unless he is born again.

NICODEMUS: How can a man who is old be born twice?

JESUS: You should not be surprised at my saying this: you must be born again.

NICODEMUS: How can this be?

JESUS: I tell you the truth, we speak of what we know, and we testify to what we have seen, but still you people do not accept our testimony. I have spoken to you of earthly things, and you don't believe; how will you believe if I talk about heavenly things? No one has gone into heaven except the one who came from heaven–the Son of Man.

*(moves closer to* NICODEMUS *and looks into his eyes)* Because God so loved the world that he gave his only Son, that anyone who believes in him will not die but will live forever. *(touching* NICODEMUS *on the arm)* This is the verdict: Light has come into the world, but men loved darkness instead because their actions were evil. Everyone who does evil *(music begins)* hates the light, and will not come into the light because they fear their evilness will be revealed. But whoever lives in the truth comes into the light, so that it may be seen plainly what he has done has been done through God.

*(*NICODEMUS *is greatly affected by these words.* JESUS *pauses a moment, and then He exits into the dark of night.* NICODEMUS *falls to his knees. As the song concludes the lights go to black.)*

# More than Just a Man

Words and Music by
CHRIS MACHEN
*Arranged by Richard Kingsmore*

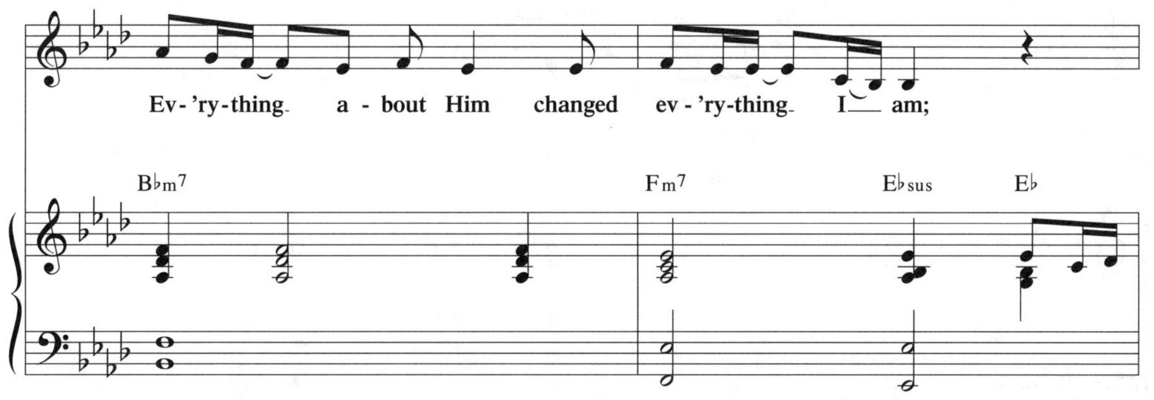
Ev-'ry-thing a-bout Him changed ev-'ry-thing I am;
B♭m7
Fm7
E♭sus
E♭

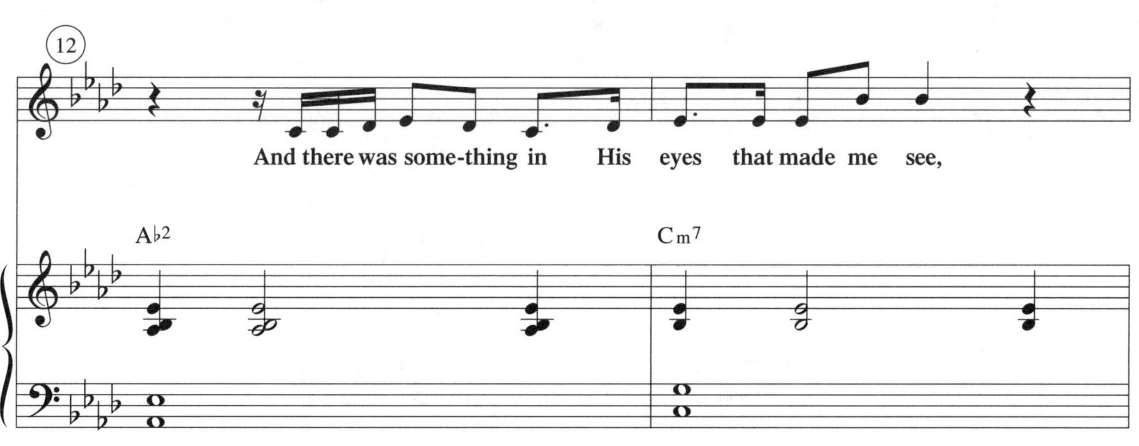
12
And there was some-thing in His eyes that made me see,
A♭2
Cm7

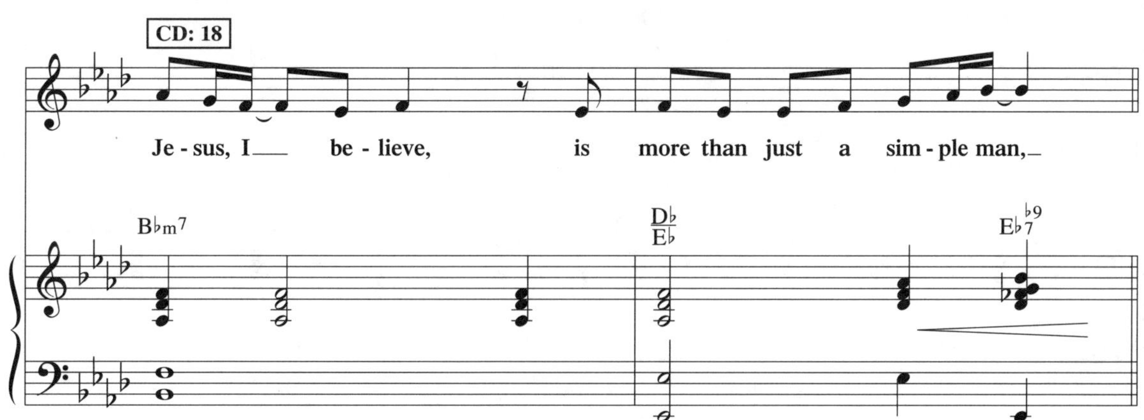
CD: 18
Je-sus, I be-lieve, is more than just a sim-ple man,
B♭m7
D♭/E♭
E♭7♭9

16

Faster ♩ = ca. 72

mf

He is more than just a man,

*Choir, 2nd time only*

mf

More than just a man,

mf

16

A♭ E♭sus E♭

mf

He is more than just an or - di-nar - y man,

Oo, ah.

B♭m7 A♭/E♭ E♭

20
He's the One who is the Sav - ior and the Christ, This
One who is the Christ.
20
A♭2
A♭/E♭
E♭
Man who changed my life is more than just a man,
E♭/D♭
D♭
B♭m7

24
CD: 21
2nd time
So much more,
More than just a man.
A♭2/C
D♭2
2nd time to Coda
(to pg. 38, meas. 37)
CD: 19
more than just a man.
D♭/E♭
A♭M7
D♭2/A♭
E♭7♭9/A♭
E♭7♭9
29
I met a Man whose voice spoke truth in-to my night,
A♭2
Cm7

Words of peace and pas-sion,
words of love and light;
Oo,
B♭m7
Fm7
E♭sus
E♭
33
And there was some-thing in His eyes and in His heart That
Oo.
33
A♭2
Cm7

CD: 20
D.S. al Coda
(to pg. 34, meas. 16)
set this man a - part, and made me want to fol-low Him.
D.S. al Coda
(to pg. 34, meas. 16)
B♭m7
D♭
E♭
E♭7 ♭9
CODA
38
So, this is how it feels to be born a-gain,
Ah,
CODA
38
A♭
G♭M9
F m

To know Je - sus as my Lord.
Je - sus as my
Cm7
Fm
Fm/E♭
42
So this is what it means to be whole a-gain,
Lord.
A♭/D♭
D♭
Cm7

CD: 81
Optional final reprise
CD: 22
And to know He's so much more.
3
He's
B♭m7
A♭
C
E♭sus
E♭
47
He is more than just a man,
so much more!
E♭sus
E♭
47
B♭

He is more than just an
More than just a man,
just an
F sus
F
C m7
51
or - di-nar - y man,
He's the One who is the
or - di-nar - y man,
Ah,
B♭
F
F
51
B♭2

Sav - ior and the Christ, This Man who changed my life,

B♭/F F F/E♭ E♭

55

He is more than just a man,

More than just a man.

E♭/F 55 B♭

43
CD: 23
He is more than just an
More than just a man, just an
F sus
F
C m7
59
or - di-nar - y man,
He's the One who is the
or - di-nar - y man,
Ah,
B♭
F
B♭2

Sav - ior and the Christ,
This Man who changed my life
B♭/F
F
F/E♭
E♭
is more than just a man,
63
So much
Cm7
B♭2/D

More than just a man,
more,
He is
E♭
B♭2/D
67
He is more than just a man,
so much
more,
so much
E♭
67
B♭2/D

more,
Je - sus is
more,
He is more than just a
man,
A♭6 9
E♭2 G
more,
He's
more,
than just a
rit.
He is more,
more than just
a
D♭ G♭
G♭M7
71
Fm7
B♭ F
Fm7
rit.

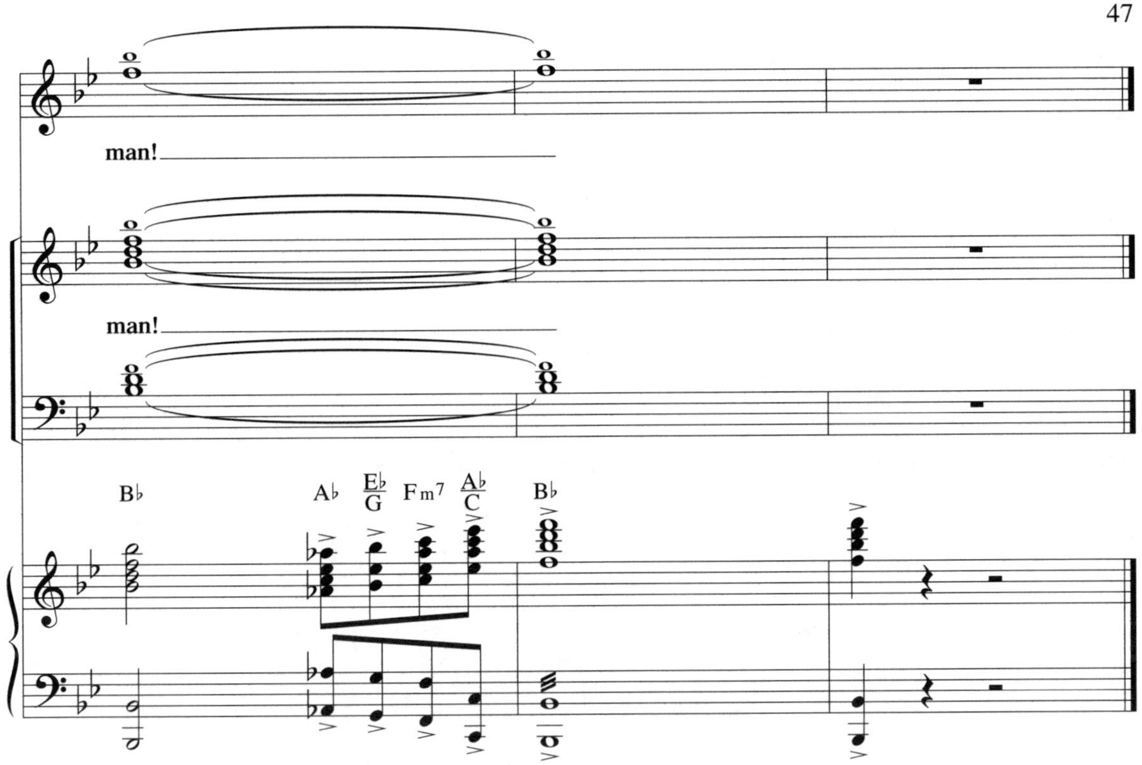

man!

man!

B♭ A♭ E♭/G Fm7 A♭/C B♭

# Underscore 3

(Call to Worship)

CHRIS MACHEN

*Arranged by Richard Kingsmore*

With intensity ♩ = ca. 68

# Scene 4

*John 7-9*

*(It is the next week.* CAIAPHAS *has convened the Sanhedrin.* JOSEPH *and* NICODEMUS *are absent. As the lights go up, the men are in the midst of their conversation. The voices start in the darkness.)*

ELIAKIM: We must take action.

PEKAH: We cannot ignore the reports coming in.

JONATHAN: He camps out at our temple polluting the minds of the people.

JORAM: He forgave that adulterous woman on the temple grounds! Who but God has the power to forgive?

ELIAKIM: Caiaphas, the crowds are getting rowdier. We need to do something.

*(Everyone present responds saying things such as, "Yes.", "We must!", etc.)*

CAIAPHAS *(quieting them and then)*: Please, gentlemen. In the past, we have dealt with false prophets. But breaking the law, and blasphemy is something we will not deal with lightly.

*(Again, the council responds with some of the men saying, "No!", "That's right.", etc.)*

CAIAPHAS *(looking around)*: Something must have detained Joseph and Nicodemus. Nevertheless, it is my opinion that we can't ignore this situation. The next time He shows His face at the temple, Jesus will be brought in for questioning.

*(The council members respond: "Yes!", "Bring Him in!", etc. During these lines,* JOSEPH *enters but purposely stays out of* CAIAPHAS*' line of sight. He is a bit distracted. The council members notice* JOSEPH *and acknowledge him first.)*

CAIAPHAS *(seeing him)*: Joseph. How nice that you could finally join us. We were just discussing Jesus, and how to deal with Him. Where is Nicodemus?

JOSEPH: I don't know. That is why I was late. I went to look for him.

*(*NICODEMUS *suddenly enters.)*

CAIAPHAS: Here you are. *(perplexed but wanting to move on)* Good. We have heard many things about Jesus this week. *(He goes on talking not realizing the independent meeting* JOSEPH *and* NICODEMUS *are having at first.)*

JOSEPH *(pulling* NICODEMUS *aside slightly, whispering)*: Where have you been? You just disappeared. What have you been doing?

NICODEMUS: I was with the Master.

JOSEPH: Who? What are you talking about…you mean Jesus?

*(*JOSEPH *and* NICODEMUS *realize everyone is waiting for them to speak.)*

CAIAPHAS *(a bit annoyed)*: Excuse me. Friends, do you have anything new to report?

JOSEPH and NICODEMUS *(at the same time)*: No.

CAIAPHAS: Good. Before you two arrived, we decided to bring Jesus in.

NICODEMUS *(quietly)*: No one ever spoke the way this man does.

CAIAPHAS: What?

NICODEMUS *(more boldly)*: No one ever spoke the way this man does.

*(There is an awkward pause again.)*

CAIAPHAS: So, you were impressed? *(looking at* NICODEMUS*)* I don't want any of you being deceived. Do you understand me?

JONATHAN: The mob that follows him knows nothing of the law. There is a curse on them.

NICODEMUS: Does our law condemn anyone without first hearing him to find out what he is doing?

PEKAH: Are you from Galilee, too? Search the ancient prophecies and you will be convinced that a prophet does not come from Galilee.

CAIAPHAS: Enough debate! Jesus is compromising the peace, and He is compromising our law. Joseph!

JOSEPH: Yes.

CAIAPHAS: I'd like you to handle this personally.

JOSEPH: I'm sure any one of the council members can…

CAIAPHAS *(cutting him off)*: But *you* are my choice. Bring Jesus to us, so we may find out who this "messiah" really is. *(*JOSEPH *nods in acceptance of his task)* You are all dismissed. *(Everyone begins to exit.)* Nicodemus, I'd like a word with you.

*(*NICODEMUS *approaches* CAIAPHAS*;* JOSEPH *looks back at them as he exits. The lights quickly fade to black.)*

# Underscore 4

(Hallelujah for the Cross)

CHRIS MACHEN
*Arranged by Richard Kingsmore*

# Scene 5

*John 8:31-36*

*(As the lights go up,* JESUS *is at the temple again preparing to speak to the crowd and His* DISCIPLES. *Some are sitting; others standing.* NICODEMUS *is present and close to* JESUS. *There are other* COUNCIL MEMBERS *amongst them. The crowd is talking loudly; some are talking to each other about* JESUS. *A small group has surrounded* JESUS, *and the* DISCIPLES *are collectively trying to watch out for Him.)*

PETER: Please! Please! Settle so he can speak.

*(A few seconds pass as the crowd hushes.* JOSEPH *enters and stands away from* JESUS *and* NICODEMUS *so as not to be seen.)*

JESUS *(as he makes his way around the crowd, first addressing his* DISCIPLES*)*: If you hold to my teaching, you are really my disciples. *(going to* NICODEMUS *and looking at him while saying:)* Then you will know the truth, and the truth will set you free.

JORAM: How can you say that we will be set free?

JESUS *(addressing everyone)*: I tell you the truth, everyone who sins is a slave to sin. Now a slave has no permanent place in the family, but a son belongs to it forever. *(music begins)* So if the Son sets you free, you will be free indeed.

*(As* POWER IN THE PRESENCE OF THE LORD *begins and the choir begins to sing,* JESUS *begins to perform miracles. He moves through the crowd. A sick child is brought to Him; a blind man; a demon-possessed woman, and He heals each of them. The* COUNCIL MEMBERS *are not happy at these events unfolding, but* JOSEPH *is very amazed and not aware of himself. Towards the end of the song,* NICODEMUS *sees* JOSEPH, *noting how preoccupied with* JESUS JOSEPH *is.)*

# Power in the Presence of the Lord

Words and Music by
CHRIS MACHEN
*Arranged by Richard Kingsmore*

9
touch His ten-der hand; He spoke with all au - thor - i - ty, the
D2/A
Bm7

peo-ple stood in awe, This kind of pow - er Je-sus had could
F♯m7
Bm7
A2/C♯

CD: 27
Choir
mf
on - ly come from God.
There is
G2
D/F♯
Esus
E
E4/2
E

15
pow - er in the pres - ence of the Lord. There is
A2 E/D DM7 E/D D A2/C♯
mf
peace and hope a - bid - ing in the Sav - ior. There is
F♯m7 Bm7 A/G G E/G♯
19
won - der as we wor - ship and a - dore. There is
D2 C♯7sus C♯7/E♯ F♯m F♯m/E

22
pow - er in the pres - ence of the Lord.
B m7
D
E
A 2
CD: 28
His
D 2
A
A 2
D 2
26
His
heart was filled with pas - sion, His words were filled with grace.
B♭2
E♭2

eyes were filled with mer - cy as He looked up-on each face.
In the
B♭2
E♭2
30
pres-ence of the Sav - ior there is hope and there is peace,
And
Cm7
Gm7
cresc.
CD: 29
all who come to Je - sus find He's all they ev - er
Cm7
B♭2/D
A♭2
E♭/G

36
need.
There is pow - er in the pres - ence of the
Fsus
F
E♭/F
B♭2
F/E♭
E♭M7
E♭
Lord.
There is peace and hope a - bid - ing in the
B♭2/D
Gm7
Cm7
40
Sav - ior.
There is won - der as we wor - ship and a -
B♭/A♭
A♭
F/A
E♭2
D7sus
D7/F♯

CD: 30

dore. There is pow - er in the pres - ence of the

Gm Gm/F Cm7 E♭/F

Lord. In His pres - ence___ there is pow - er in His

44

B♭ B♭/A♭ G♭ A♭ G♭/A♭ Fm/A♭ E♭m/A♭

name. In His pres - ence___ we will nev - er be the

D♭2 E♭m7 Fm

48
CD: 31
same,
nev - er be the same,
Cm
F
Dm
G
Em
G
F
G
molto rit.
51
a tempo
There is pow - er in the pres - ence of the
G sus
C2
F M7
F
Lord.
There is peace and hope a - bid - ing in the
C2
E
A m7
D m7

55
Sav - ior. There is won - der as we wor - ship and a -
C/B♭ B♭ G/B F2 E7sus E7/G♯
CD: 32
dore. There is pow - er in the pres - ence of the
Am Am/G Dm7 F/G
ff
59
Lord. There is pow - er in the pres - ence of the
C D♭ A♭/G♭ G♭M7 A♭/G♭ G♭
cresc.
8vb

Lord.
There is peace and hope a-bid - ing in the
D♭2/F
B♭m7
E♭m7
63
Sav - ior.
There is won-der as we wor-ship and a-
D♭/C♭
C♭
A♭/C
G♭2
F7sus
F7/A
dore.
There is pow - er in the pres - ence of the
B♭m
B♭m/A♭
E♭m7
G♭/A♭

67
Lord.
There is
pow - er,
there is
B♭m
B♭m
A♭
E♭m7
pow - er,
There is
pow - er
in
the
D♭2
F
G♭
G♭6
G♭M7
G♭
CD: 33
pres - ence
of
the
Lord.
There is
G♭M7
A♭
E♭m
A♭
G♭
A♭
A♭
D♭
C♭
G♭
B♭

73
pow - er in the pres - ence
A
A♭m7
F♯m7
A/E
of the Lord, the
G♭m2/E♭
G♭m/E♭
G♭m/A♭
D♭
C♭
77
rit.
fff
Lord, the Lord!
A
G♭m/A♭
D♭
8vb

*(As* POWER IN THE PRESENCE OF THE LORD *ends,* NICODEMUS *approaches* JOSEPH.*)*

NICODEMUS: Joseph.

JOSEPH *(a bit startled)*: Nicodemus. I…Jesus…He is…

NICODEMUS: Yes, He is. *(There is an uncomfortable silence.)*

JOSEPH *(back in reality)*: I was…I am concerned about how you seem to feel about Jesus.

NICODEMUS: Yes, I believe. I believe He is the Messiah. And you?

JOSEPH *(unsure)*: I don't. I can't. Oh, I don't know…

NICODEMUS: This man is the Son of God, Joseph.

JOSEPH: You heard what Caiaphas and the others said.

NICODEMUS *(gently)*: Yes, and you have heard what Jesus says.

JOSEPH: But we uphold the law. The facts do not add up.

NICODEMUS: You mean *our* facts? This isn't about those supposed facts. This is about truth.

JOSEPH: What?

NICODEMUS: I talked to Jesus. He gave me a new life.

JOSEPH: Don't we have a great life?

NICODEMUS: Joseph, you don't have to be afraid. *(touching* JOSEPH'S *arm)* Listen to what Jesus said to me: "Because God so loved the world he gave his only Son, that anyone who believes in him will not die but will live forever."

JOSEPH *(moving away)*: It doesn't make sense.

NICODEMUS: Every one of us–for all of our lives–has had an idea of what "messiah" means. Now He's here, and He's so much more. Jesus has redefined what we thought. Now I know a peace that I've never known before. *(carefully)* And I know you feel it too.

JOSEPH: I have to go back to the council. I have a duty. Caiaphas…he's expecting me…

NICODEMUS: Joseph, don't do it.

JOSEPH *(looking at him)*: I have to.

*(He exits leaving* NICODEMUS *who watches as he goes and lights fade to black. Music begins.)*

# Scene 6

*(As lights come up, we are at another Sanhedrin meeting.* JOSEPH *is present,* NICODEMUS *is not. The* COUNCIL *has decided to do what needs to be done to Jesus.)*

## Who Is This Man?

Part 2

Words and Music by
CHRIS MACHEN
*Arranged by Richard Kingsmore*

dore Him. They're hang-ing on ev - 'ry word.
G2 Dm7 C/F G2
13
Solo 3
mf
Some-how we have to con - trol Him Or soon we will nev - er be heard
Dm7 C/F G2 Dm7 C/F
CD: 35
from a-gain.
Men join solo
mf
This
G2 G/A Am Am9
19
Je - sus! This Je - sus!
FM7 G

23
CD: 36
Who is this man?
G2
Am
F2
G
27
Solo 3
mf
Who is this Man?
C2
Why can't they see
No pro- phet hails
from Gal - i - lee?
D/C
C2
D/C
31
Solo 1
mf
All of the peo - ple are turn - ing.
They're call-ing Him Mas - ter and Lord.
Dm7
C/F
G2
Dm7
C/F

Solo 4
35
mf
Some-thing has got to be done here Be -
G2
Dm7
C/F
G2
CD: 37
fore we're com-plete - ly ig-nored for all time.
Dm7
C/F
G2
G/A
Am
Men join solo
mf
41
This Je - sus! This
Am9
FM7
45
Je - sus! This Je - sus!
G
FM7

Who is this

FM7 G2

(49)

Man?

G/A Am G/A Am G/A Am G/A

G/A Am

(53) Am2 CD: 38

Slower ♩ = ca. 76

*mp*

**Caiaphas begins*

Am2 B♭/A (57) E♭/A

*CAIAPHAS: Gentlemen, listen to reason!
There's something you're failing to see:
Before us there lies a soloution.
This Jesus has given it to me!

CD: 39
Solo (Caiaphas)
mf
63
Mes - si - ahs have ris - en be -
Gm
A
E 7
A
Am2
mf
fore Him Who've led man-y peo-ple a - stray.
FM7
Am2
FM7
67
Let's not be too harsh on these ig - no-rant sheep. It's the
Dm7
Am2
CD: 40
Shep - herd who stands in our way.
Dm11
Dm7
E 7

All Sanhedrin Men
71
mf
Who is this Man? Why do we care? We'll blot out His mem - 'ry like He
Am7
D/A
Am7
nev - er was there.
75
f
One man will die for the na - tion!
D/A
Dm7
C/F
G2
f
One man will die in their place!
Dm7
C/F
G
79
One man will die for the
Dm7
C/F
peo - ple!
One man will die in dis-grace, one for
G2
Dm7
C/F
G

85
all.
This Je - sus!
Am G/A Am Am/G G/F F
This Je - sus!
This
G/F F Am/G G Am/G G
89
ff
Je - sus, He is the Man!
Dm6/A Am E/G♯ E A4/2 Am E/A
ff
rit.
A4/2 Am E/A A4/2 Am E/A A2(no 3)
rit.

CAIAPHAS: Where is Nicodemus? *(looking around)* Joseph? Do you know where he is?

JOSEPH: He follows Jesus now.

CAIAPHAS: Well. I'm afraid he is immediately relieved of his duties to this council. And Jesus? *(angry)* Where is Jesus?

JOSEPH: I…I couldn't…

CAIAPHAS: You couldn't what? *(pause)* Joseph! Why isn't Jesus here?

*(There is no answer.)*

JOSEPH *(quietly)*: No one ever spoke the way this man does.

CAIAPHAS: What?

JOSEPH: Caiaphas, I can't shake His words. His teachings they are…

CAIAPHAS: Joseph. What are you doing?

JOSEPH: I'm sorry. I'm not sure what to do.

CAIAPHAS *(upset)*: I don't know what has happened to you. But hear this: the man you are thinking about following is a dead man. Do you understand? *(pause)* Do you?

JOSEPH: Caiaphas, please…

Caiaphas: The council has made a decision. If you don't stand with us, then you stand outside of us.

*(*JOSEPH *looks around at the council. He shakes his head. He cannot be a part of their condemnation. He runs out of the meeting.)*

CAIAPHAS *(forgetting himself for a minute)*: Joseph! *(becoming quickly aware of his position and addressing the council)* So, we lose two members of the council in one night. This Jesus will not evade us forever. That I promise. Let's see this blasphemer fall!

*(The* COUNCIL MEMBERS *lift their voices in agreement as the lights go to black.)*

# Scene 7

*Luke 19:37-40, 45-48*

*(Lights go up and music begins. A crowd enters with* JESUS *as He comes into Jerusalem. The people are waving palm branches–a sign that He is king.)*

## Blessed Is He

Words and Music by
CHRIS MACHEN and MARK BOVEE
*Arranged by Richard Kingsmore*

mf
13
Ho - san - na to the Son of Da - vid, Ho - san - na in the high -
mf
13 Am G Am Dm
17
est. Ho - san - na to the Son of Da - vid, Ho -
E 17 Am G Am
CD: 42
21
f
san - na in the high - est. Bless - ed is He who
f
Am E Am G/B A/C♯ Dm7 G
21
f

comes in the name, Who comes in the name of the Lord;
G Am Am/G F G Am G
25
Bless-ed is He who comes in the name, Who comes in the name of the Lord,
1 CD: 43
Dm7 G Am Am/G
1 F Esus E
(to pg. 74, meas. 21)
the Lord! comes in the name of the Lord!
2
30
Am G/B A/C♯ A
F Dm Esus

CD: 44
mf
Ho -
mf
E
Am
E
Am
34
san - na, ho - san - na, Ho - san-na in the high -
34
Am
G
A 4 2
Am
Dm
mf
38
est, Ho - san - na, ho - san - na, Ho -
E
38
Am
G
A 4 2
Am

san - na in the high - est.

Am E A(no 3)

decresc.

43 Am *ELIAKIM: Teacher, rebuke your…* CD: 45 E

*mp*

*(The group stops as someone yells at* JESUS *for the commotion He's causing.)*

ELIAKIM: Teacher, rebuke your disciples!

JESUS: I tell you, if they keep quiet, the stones will cry out.

*(The crowd continues to journey with* JESUS.*)*

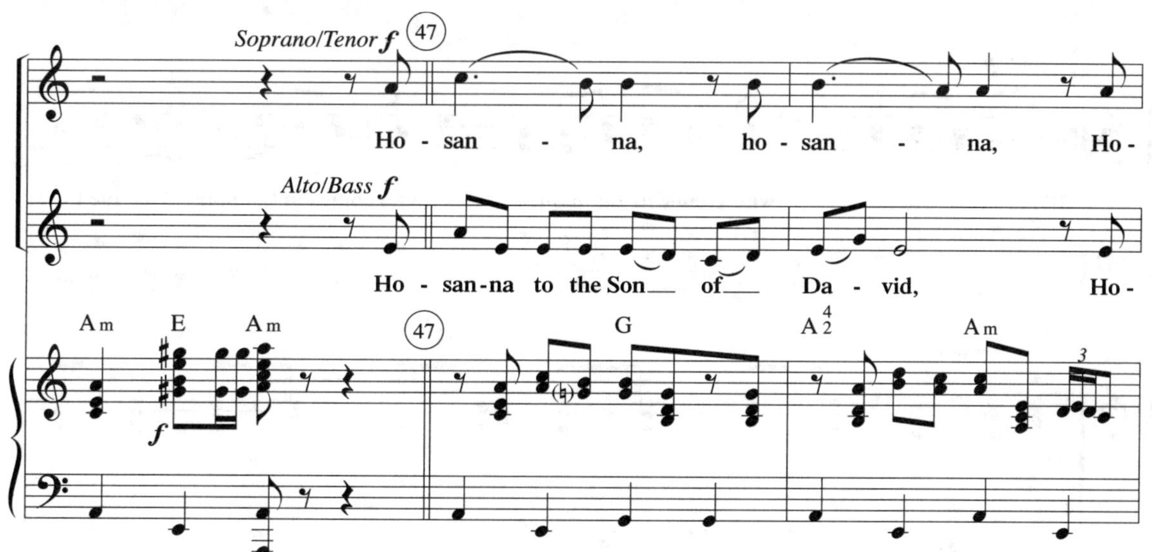

51
san-na in the high - est, Ho - san - na, ho -
san-na in the high - est. Ho - san-na to the Son of
Am Dm E Am G
CD: 46
san - na, Ho - san-na in the high - est.
Da - vid, Ho - san-na in the high - est.
A4 2 Am E Am G/B A/C♯
55 S.A.T.B.
Bless-ed is He who comes in the name, Who comes in the name of the Lord;
Dm7 G Am Am/G F G

59
Bless-ed is He who comes in the name, Who
Am G Dm7 G Am Am/G
1
(to pg. 78, meas. 55)
2
comes in the name of the Lord, the Lord! comes in the name of the Lord!
F Esus E Am G/B A/C♯ A F Dm
64
Esus E N.C.

Am
68
Gm/A
CD: 47
B♭/A
Am7
Am
mp

JESUS: It is written…
Gm/A
B♭/A
Am7
Am
72
Gm/A
B♭/A

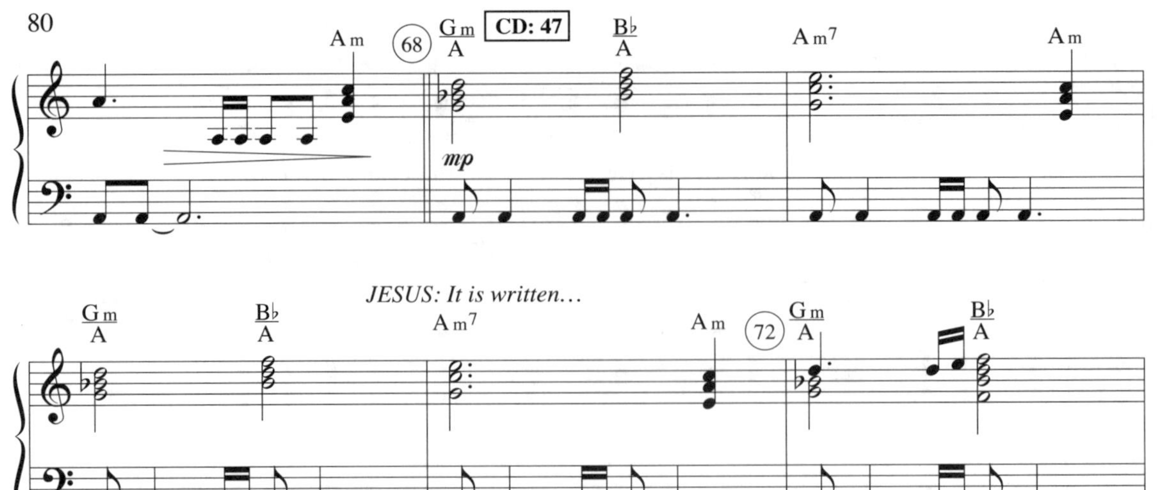

*(*JESUS *arrives outside of the temple to drive out the vendors.)*

JESUS: It is written, "My house will be a house of prayer," but you have made it a "den of robbers."

*(*JESUS *begins tearing apart the vendor's belongings. The disciples and the crowd are ecstatic at* JESUS' *actions and they respond by cheering and continue to sing* BLESSED IS HE. *The Sanhedrin leaders look on with disdain, but they don't know what to do.)*

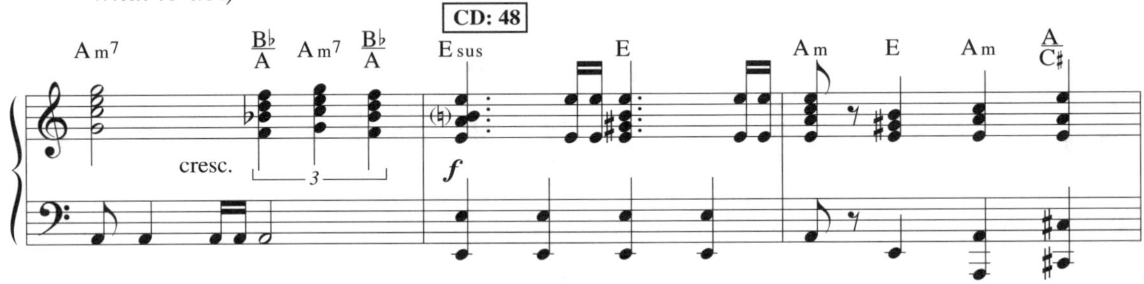

CD: 48
Am7
B♭/A
Am7
B♭/A
Esus
E
Am
E
Am
A/C♯
cresc.
3
f

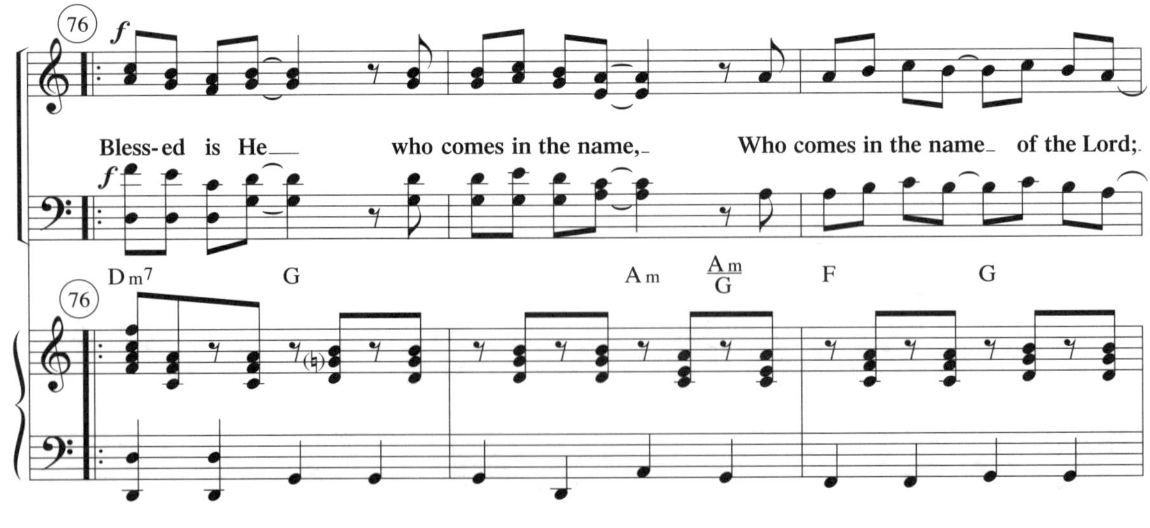

76
f
Bless-ed is He who comes in the name, Who comes in the name of the Lord;
Dm7
G
Am
Am/G
F
G

80
Bless- ed is He
who comes in the name,
Who
Am
G
Dm7
Am
Am
G
1
(to pg. 80, meas. 76)
2
comes in the name
of the Lord,
the
Lord!
comes in the name
of the Lord!
F
Esus
E
G
B
A
C♯
A
Dm
86
Who
comes
in
the
name,
who
Esus
E
F
G
F

90
comes in the name, Who comes in the
F G/F F G Am/G G FM7 G/F F
name of the Lord.
Esus E Am Am/G
94
Bless - ed is He who comes in the
F G Am Am/G F

# Scene 8

*John 18-19; Luke 22-23*

*(*JOSEPH *is praying alone in a garden outside of Jerusalem. He has been gone for a week. The lights are dim. Music begins.)*

JOSEPH: Father, I have done a lot of searching this week. I've tried to sort out what I've seen Jesus do. To understand the radical way he lives. Lord, I have followed you all my life, but I've been confused lately. You know how I've struggled in my mind and in my heart. But I believe Jesus is your Son.

There are things I have done in wrong ways. I have sought after the glory of men and not you. Forgive me, Lord. I want to honor you and not myself. I want to be the man you want me to be. Your Son's words have pierced my heart, and I want a new life–starting now.

*(*NICODEMUS *enters as* JOSEPH *begins to sing but* JOSEPH *doesn't see him at first.* NICODEMUS *joins in singing at measure 25.)*

# More than Just a Man

Part 2

*with*

*Underscore 5 (Joseph's Prayer)

Words and Music by
CHRIS MACHEN
*Arranged by Richard Kingsmore*

14
Solo (Joseph)
mp
So, this is how it
C/D
D sus
D/C
C sus/C
B♭M7
cresc.
mp
feels to be born a-gain,
to know Je - sus as my
Dm
Am7
Lord.
18
mf
So, this is what it
Dm
Dm/C
B♭M7
mf
means to be whole a-gain,
and to know He's so much
3
Am7
Gm7
F/A

CD: 52
cresc.
more.
C sus
C
C sus
C
cresc.
23
f
He is more than just a man,
G
D sus
D
f
Nicodemus joins
mel.
He is more than just an or - di-nar - y man.
A m7
G
D
D
27
CD: 53
He's the One who is the Sav - ior and the Christ. This
G 2
G
D
D

Man who changed my life,

More than just a

D/C C C/D

31

He is more than just a man,

man. More than just a

31 G2 Dsus D

He is more than just an or - di-nar - y man.
man, just an or - di-nar - y man,
Am7
G/D
D
35
He's the One who is the Sav - ior and the Christ, This
Ah
35
G2
G/D
D

Joseph
Man who changed my life is more than just a man.
D/C
C
Am7
39
CD: 54
Nicodemus
He's more than just a man,
So much more, He is
G2/B
C
G2/B

43

*Both*

He is more than just a man, So much

more, So much

C 43 G2/B

more.

Jesus is

more, He is more than just a man,

F9/6 C2/E

47
rit.
more, He's more, than just a
He is more, more than just a
Bb/Eb
EbM7
47
Dm7
G/D
Dm7
rit.
molto rit.
man!
man!
G
F
C/E
Dm7
F/A
G
molto rit.

*(For a moment after the song, neither man knows what to say or do.* NICODEMUS *finally hugs* JOSEPH. *And the two men have a hearty reunion.)*

NICODEMUS: I've been searching for you all week.

JOSEPH: I didn't know what to do. I had to get away–to clear my head. To talk to God. *(pause)* I'm sorry about before. I was confused, but now I know what you said is true–Jesus is the Messiah.

NICODEMUS: All of heaven is rejoicing for you my friend! *(pause)* It has been a long week.

JOSEPH: I hope Caiaphas has cooled off by now. I've known him all my life, he can be so…

NICODEMUS *(cutting him off)*: Joseph. *(pause)* Caiaphas has had Jesus arrested.

JOSEPH: What?

NICODEMUS: Apparently one of Jesus' followers led the council straight to Him.

JOSEPH: Betrayed by a friend.

NICODEMUS: The council tried Jesus–if you could call it a trial. They beat Him and took Him to Pilate.

JOSEPH: To Pilate?

NICODEMUS: Since we have no law that allows us…

JOSEPH *(finishing the thought)*: To execute a man…Caiaphas, no!

NICODEMUS: Pilate wouldn't do it. He sent everyone to Herod. But Herod wouldn't do it either. Now Jesus is back in Pilate's hands. The council and the crowd are demanding a crucifixion.

JOSEPH *(greatly saddened)*: Caiaphas, why such extremes?

NICODEMUS: We have to hurry Joseph. We have to try and stop them.

JOSEPH: Yes. Let's go!

*(They exit. Lights go to black.)*

# Scene 9

*John 19; Luke 23*

*(In the darkness, the pounding of nails is heard. After it stops, music begins. As the lights come up,* JESUS *is laying on the cross, which is still on the ground. A crowd is nearby including* MARY, Mother of Jesus *and other women, some of the* COUNCIL MEMBERS *and* CAIAPHAS. *The cross is raised by* ROMAN SOLDIERS *at the chorus, and a badly beaten* JESUS *hangs on the wood. During the second verse, His breathing becomes more labored.* MARY *gets closer to her Son and reaches up to Him.* JOSEPH *and* NICODEMUS *arrive realizing at once that they are too late. When* JOSEPH *and* CAIAPHAS *see one another, they stare in each other's eyes for a moment.* CAIAPHAS *looks indignant but finally leaves the scene.* JOSEPH *and* NICODEMUS *move closer to the cross as* JESUS *takes His final breaths. At the end of the song,* JESUS *takes a final breath in, and His head drops. The song concludes and everyone silently beholds the crucified Christ. The women weep. The lights fade with a special on only* JESUS. *No one moves, except for* JOSEPH. *A light comes up on* JOSEPH, *as he stands. He knows what to do. He takes a look at* JESUS *and then quickly leaves the scene. The light on* JESUS *fades slowly to black.)*

# Hallelujah for the Cross

Words and Music by
CHRIS MACHEN
*Arranged by Richard Kingsmore*

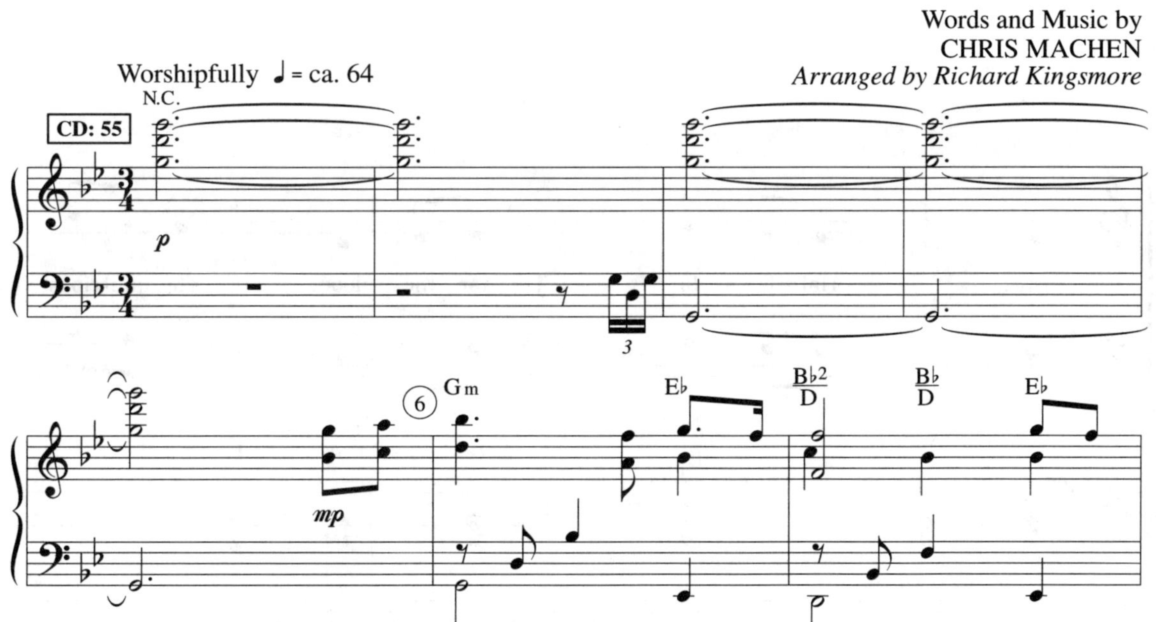

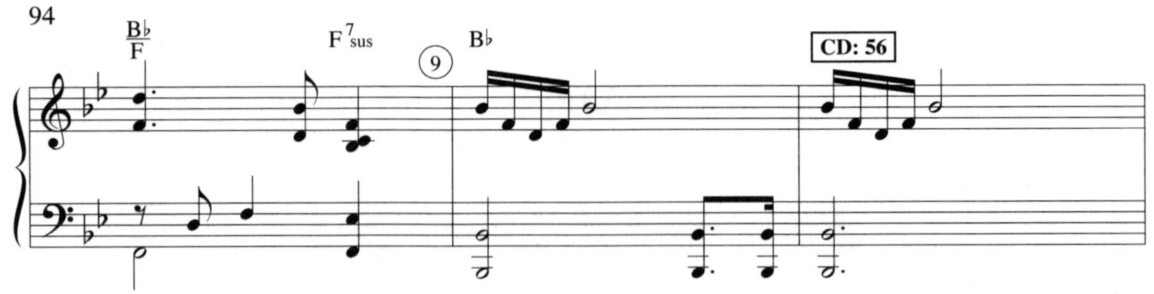
B♭
F
F 7 sus
9
B♭
CD: 56

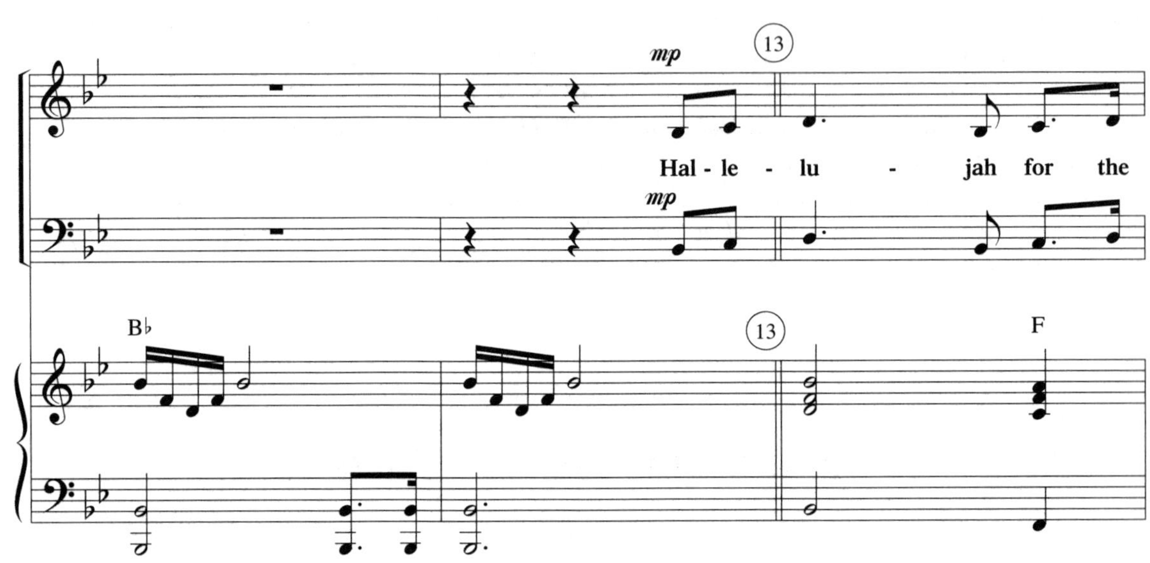
mp
13
Hal - le - lu - jah for the
mp
B♭
13
F

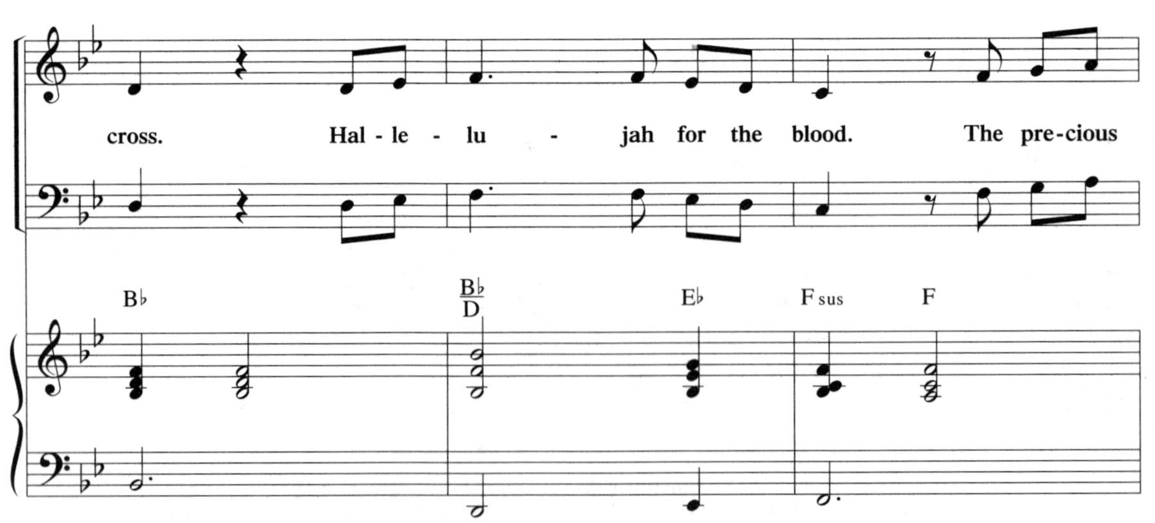
cross. Hal - le - lu - jah for the blood. The pre-cious
B♭
B♭
D
E♭
F sus
F

17
blood that flowed from Him would cleanse the deep - est stain of
17
Gm
E♭
B♭2/D
B♭/C
F2
sin, would cleanse the deep - est stain of sin.
22
Gm7
B♭/F
B♭/C
F sus
F
22
B♭(no 3)
E♭/B♭
CD: 57
mf
Hal - le -
B♭(no 3)
E♭/B♭
B♭(no 3)

26
lu - jah for the cross. Hal - le - lu - jah for the
mf
Hal - le - lu -
26
B♭
F
B♭
B♭/D
E♭
mf
crown. The crown of thorns would pierce and
jah. The thorns would pierce and
30
F sus
F
30
G m
E♭
shed, The drops of mer - cy from His head, the drops of
shed,
B♭2/D
B♭/C
F2
G m7
B♭/F

34
CD: 58
cresc.
f
mer - cy from His head. Hal - le -
B♭/C F sus F B♭sus B♭ B♭2 B♭
37
lu - jah, my soul sings, On Cal-v'ry's cross there hung the
B♭ C m7 B♭2/D E♭ B♭/F E♭/G G m7
41
King. And for my sin He suf-fered loss, I will re -
F/A F F sus/E♭ F/E♭ B♭/D E♭ B♭/F C m7

45
joice in Je - sus' cross. Hal - le - lu - jah for the
B♭/F
F
Gm
Gm/F
45
E♭
F sus
CD: 59
cross.
B♭(no 3)
E♭/B♭
B♭(no 3)
C(no 3)
F/C
50
mf
Hal - le - lu - jah for the cross. Hal - le -
C(no 3)
50
C
G
C
Dm7
mf

mf
54
Those wounds the ev - i - dence of
lu - jah for His wounds.
C/E F Gsus G
Am F
grace, For there my Sav - ior took my place, for there my
C2/E C/D G2 Am7 C/G
58 CD: 60
cresc.
f
Sav - ior took my place. Hal - le -
C/D Gsus G C(no 3)

61
lu - jah for the cross. Hal - le - lu - jah for the
61
D A D Em7 D/F♯ G
f
Lamb, the Lamb of God for sin-ners slain
65
would bear my
Asus A Bm G D2/F♯
suf-f'ring and my shame,
would bear my suf - f'ring and my
69
CD: 61
D/E A2 Bm7 D/A D/E Asus

cresc.
shame.
ff
72
Hal - le - lu - jah, my soul
cresc.
ff
D sus
D
72
G
D
cresc.
ff
sings, On Cal-v'ry's cross there hung the King. And for my
D
G
D
G2
D
D
A
G
B
G2
B
A
C♯
E sus
B
A
A sus
G
A
G
76
sin He suf-fered loss, I will re - joice in Je - sus'
76
D
F♯
G
D
A
E m7
D
A
A sus
A

80
CD: 62
cross. Hal - le - lu - jah for the cross.
Bm
Bm/A
A
Em7
Asus
A
Dsus
D
Dsus
D
83
Hal - le - lu - jah, my soul
E♭
Fm7
sings, On Cal - v'ry's cross there hung the King. And for my
E♭2/G
A♭
E♭/B♭
A♭/C
B♭/D
Fsus/C
B♭
B♭sus/A♭
B♭/A♭

CD: 63
87
sin He suf - fered loss,
E♭/G
A♭
E♭/B♭
B♭
C m
G m7
92
mp
I will re -
I will re - joice in Je - sus' cross.
C m N.C.
E♭2/B♭
E♭/B♭
A♭M9
96
joice in Je - sus' cross.
I will re - joice in Je - sus'
E♭/G
E♭2/G
F m9
E♭2/B♭
A♭M7/B♭
B♭

# Underscore 6

(More than Just a Man)

CHRIS MACHEN
*Arranged by Richard Kingsmore*

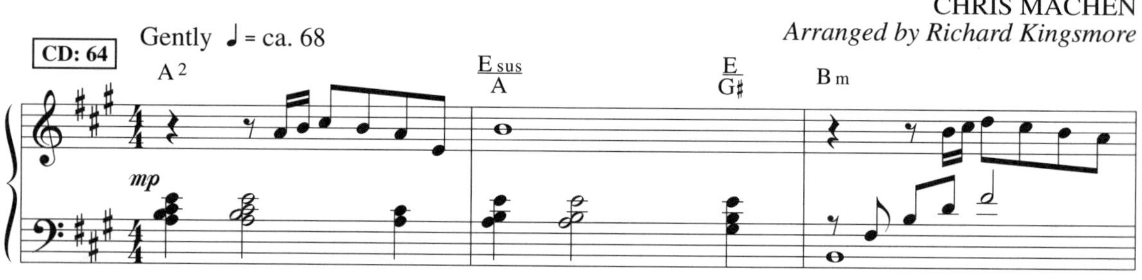

# Scene 10

*Luke 23:50-53*

*(*JOSEPH *enters, distraught. He has the linen for wrapping* JESUS*' body in his hands. He speaks to the audience as if they are Pilate. He walks cautiously down center stage. Lights are low.)*

JOSEPH *(speaking quietly)*: Sir, I am Joseph from Arimathea. I have come…*(he pauses; then speaks more boldly)* I am Joseph. I have come to ask for the body of Jesus–crucified today. *(pause)* I have a tomb for His body. I want to take it there. *(pause)* Yes, it is my tomb. *(pause)* It doesn't matter what happens now. I only want to care for…for…my Savior and Lord. *(He looks out; waiting for Pilate's answer; unsure of what his last statement will cause Pilate to do. Finally, the answer is "yes," and we see this on* JOSEPH'S *face. He turns to leave and then turns back.)* Thank you.

*(He exits. Lights to blackout.)*

# Scene 11

*John 19:38-42*

*(Lights come up on the crucifixion scene.* JESUS*' body is in* MARY*'s arms.* JOSEPH, NICODEMUS *and a few others stand around the scene. She holds him and sings.)*

## Beautiful Hands

(Part 2)

Words and Music by
CHRIS MACHEN
*Arranged by Richard Kingsmore*

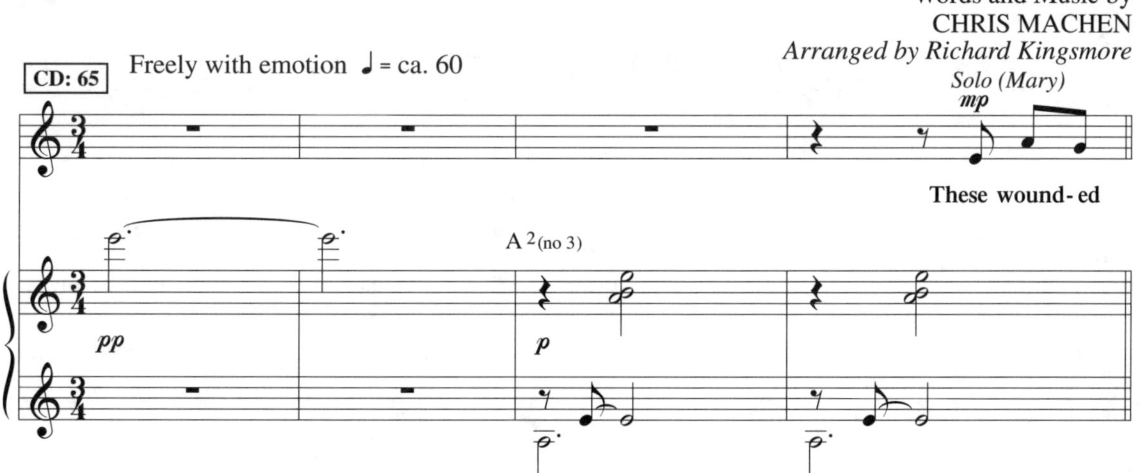

5
hands, hands of my Child, Hands pierced and
A 2(no 3)
Em11
A
mp
9
scarred, the hands of love; So much my heart can't un - der -
G 2(no 3)
F 2(no 3)
A 2(no 3)
CD: 66
rit.
stand, How beau - ti - ful these wound - ed
A 2(no 3)
D m9
E sus
rit.
a tempo
14
mf
hands. Beau - ti - ful hands once touched my
A 2(no 3)
A m2
a tempo
mf

face, beau - ti - ful hands I once em -
Am
Am/G
F2
braced, Beau - ti - ful Child, Your moth - er's
18
G2
Dm
here, And I will al - ways hold You near, I will
Am
DM9
G2
Am
22
CD: 67
rit.
a tempo
al - ways hold You near. Beau - ti - ful
Dm9
G2
A 4 2
Am
rit.
a tempo

25
hands once touched my face, beau - ti - ful hands I once em -
Ladies
mf
Oo
25 Am2 Am Am/G F2
braced, Beau - ti - ful Child, Your moth - er's here, And I will
29
Ah
G2 29 Dm Am
decresc. 32
al - ways hold You, I will
p
decresc.
Hold You,
Dm9 G2 32 A 4 2 Am

*(During* UNDERSCORE 7 (NEAR THE CROSS)*,* JOSEPH *and* NICODEMUS *take* JESUS*' broken body and lay it perpendicular to the downstage edge of the stage.* MARY MAGDALENE *and a few other women bring them the spices and aloes for the body.* JOSEPH *and* NICODEMUS *wrap the body with the spices and the linen. When they finish,* MARY, Mother of Jesus *says a final good-bye, and the two men take* JESUS*' body off stage for burial.)*

# Underscore 7

(Near the Cross)

WILLIAM H. DOANE
*Arranged by Richard Kingsmore*

With feeling ♩ = ca. 60

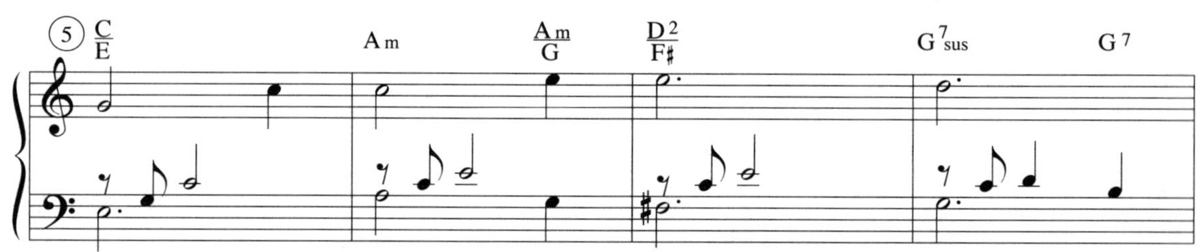


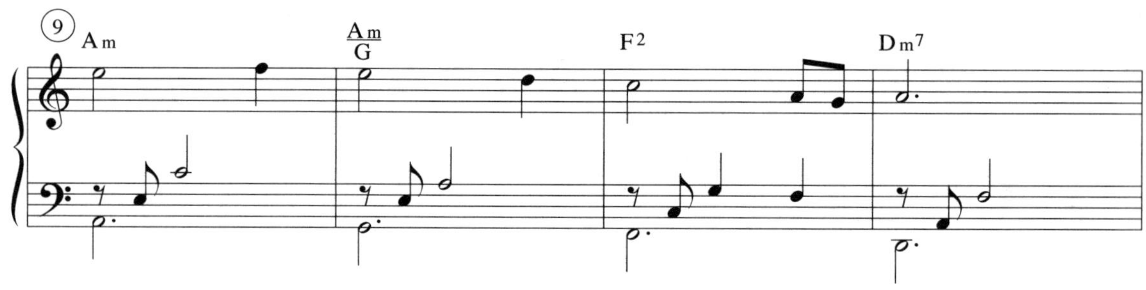

17
C2
C2/E
F
D4/2
Dm
21
Em7
Am7
Am7/G
D2/F♯
Fm6
25
E7sus
E/G♯
Am
F2
29
C2/E
Am
Dm7/G
G
Am
33
C2/G
G
Dm7/A
G/B
C

# Scene 12

*John 20; Luke 24*

*(Lights come up bright on an empty stage. After a few moments,* NICODEMUS *and* JOSEPH *enter; grieving together.)*

NICODEMUS: What now my friend?

JOSEPH: I don't know. We've wandered around since we buried Jesus. Where do we go?

*(They rest in the silence. Several seconds pass and then* MARY MAGDALENE *and* JOANNA *enter the scene.* MARY MAGDALENE *carries the cloth that* JESUS *was buried in. The women are very happy, on their way to tell others what they have witnessed.)*

MARY MAGDALENE *(seeing* NICODEMUS *and* JOSEPH*)*: Look! Those are the men who buried our Lord.

*(They approach the men.)*

MARY MAGDALENE: Excuse me. We don't mean to disturb you. Is it Joseph?

JOSEPH: Yes.

MARY MAGDALENE: We've just come from your tomb.

JOSEPH: Where Jesus is?

MARY MAGDALENE: Yes. *(excited)* I mean no, it's empty. Jesus is gone.

NICODEMUS: His body has been taken?

JOANNA: No. He has risen from the dead. *(looking at* MARY*)* Just like He told us.

JOSEPH: How do you know?

MARY MAGDALENE: I saw Him. He said my name. *(she hands the cloth to* JOSEPH*)* We have to go and tell the others!

*(The women exit, leaving the men bewildered and surprised. Music begins. They look at each other for a moment and then hug one another victoriously saying things like, "Praise God!" and then they exit quickly.)*

# Christ the Lord Is Risen Today

Segue to "That's Why I Believe"

# That's Why I Believe

Words and Music by
CHRIS and DIANE MACHEN
*Arranged by Richard Kingsmore*

Driving, with energy ♩ = ca. 106

CD: 70

A♭2

*f*

A♭2/F

5

A♭2/B♭

E♭sus

C♭2

G♭/B♭

9

A♭2

CD: 71

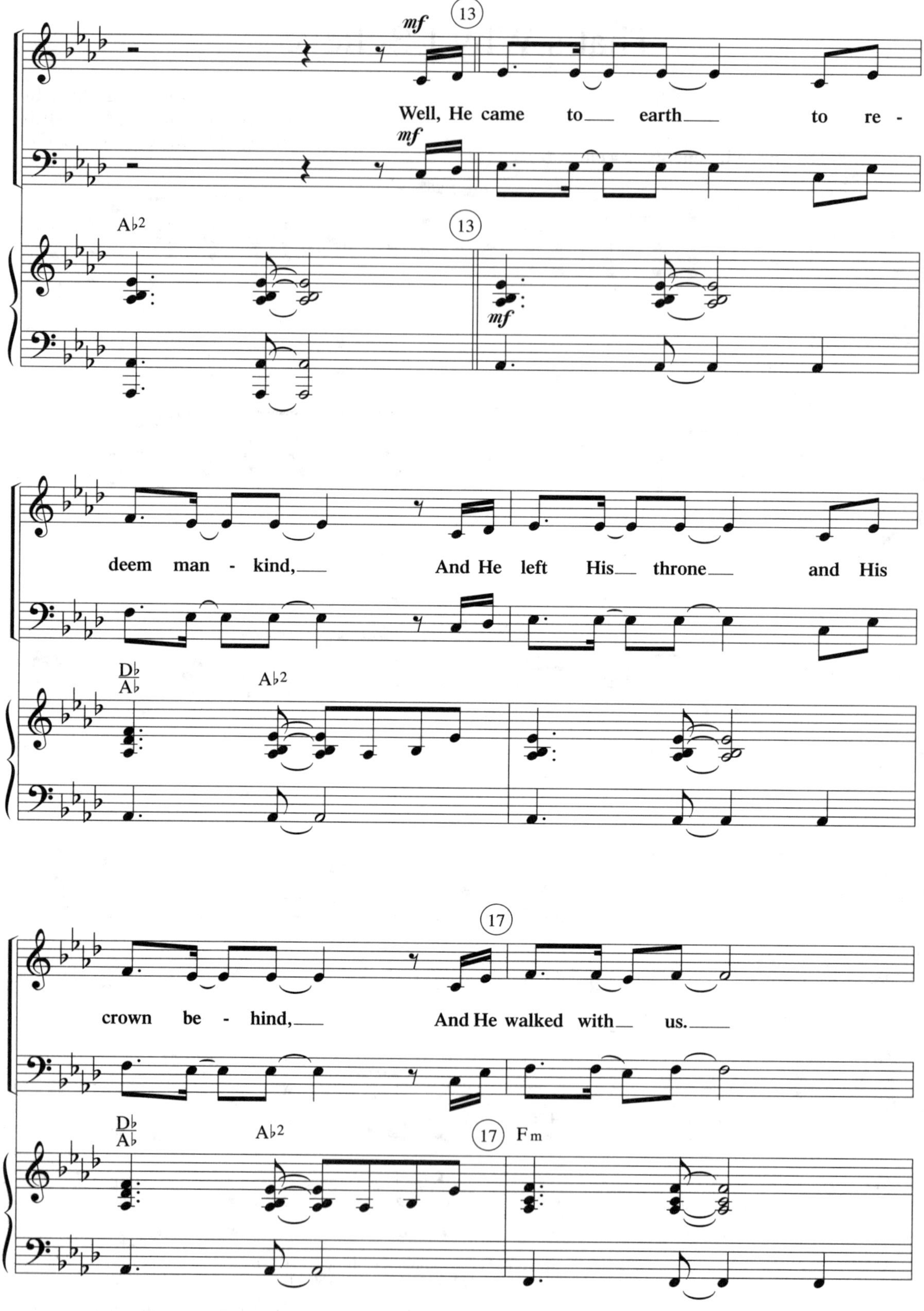

*mf* (13)

Well, He came to earth to re -

*mf*

A♭2 (13)

*mf*

deem man - kind, And He left His throne and His

D♭/A♭ A♭2

CD: 72
That's why I be-lieve.
D♭2
A♭2
21
He healed the lame, made the
A♭sus
A♭
A♭2
21
A♭
A♭2
blind to see,
And He touched the soul, set the
D♭
A♭
A♭2

25
cap - tive free,
He was God with us.
D♭/A♭
A♭2
25
Fm

That's why I be-lieve.
D♭2
A♭2

29
But the rea - son Je - sus came
D♭2/A♭
A♭
A♭2
29
E♭

CD: 73
cresc.
was to take my cross, my shame.
cresc.
E♭
cresc.
E♭/G
f
33
That's why I be-lieve.
f
E♭
33
A♭2
f
That's why I be-lieve.
A♭2
A♭2/F
Fm7

37
The Lord laid down His life,
A♭2
F
Fm7
37
B♭m
a will - ing sac - ri - fice.
E♭2
E♭
A♭2
41
Tho' they bur - ied Him that day,
A♭
41
A♭2

He rose up from the grave.
A♭2
A♭2/F
Fm7
45
That's why I believe,
A♭2/F
Fm7
45
B♭m
CD: 74
that's why I believe.
E♭sus
A♭2

49
mf
Well, He left the earth to as-
A♭2
A2
cend His throne, And the an - gels sang and the
Oo
D
A
glo - ry shone,
the King shall reign.
53
F♯m

CD: 75
That's why I be-lieve.
But
D2
A2
A
A2
57
soon He'll come to bring us home, the
57
A
A2
D/A
A2
Lamb was slain to re-deem His own, and
A2
D/A
A2

61
we shall rise.
That's why I believe.
61
F♯m
D
65
We will be for - ev - er - more
A2
D2
A
A
A2
65
E
CD: 76
cresc.
in the pres - ence of the Lord.
cresc.
E
cresc.
E
G♯

(69)

*f*

That's why I be-lieve.

E

A2

*f*

That's why I be-lieve.

A2

A2/F♯

F♯m7

(73)

The Lord laid down His life,

A2/F♯

F♯m7

Bm

a will - ing sac - ri - fice.
E2
E
A2
Tho' they bur - ied Him that day,
77
A
77
A2
He rose up from the grave.
A2
A2
F♯
F♯m7

81
That's why I be - lieve, __
A2
F♯
F♯m7
81
B m
CD: 77
that's why I be - lieve. __
E sus
A2
85
There's a bar - ren __ cross __ and an emp - ty grave, __
A2
85
D
B m7

89
And they tes - ti - fy
A2
F♯m
CD: 78
to the love He gave.
D
E sus
cresc.
ff
93
That's why I be-lieve.
F sus
N.C.

That's why I be-lieve.
97
The Lord laid down His life,
97
CD: 79
a will - ing sac - ri-fice.

101
Tho' they bur-ied Him that day,
F sus
ff
101
B♭2
He rose up from the grave.
B♭2
B♭2
G
G m7
105
That's why I be-lieve,
B♭2
G
G m7
105
C m

CD: 80
2nd time
that's why I be-lieve.
F sus
B♭2
1
(to pg. 130, meas. 101)
2
Tho' they bur-ied Him that
That's why I be-lieve,
1
(to pg. 130, meas. 101)
B♭
2
F/G
G m7
110
that's why I be-lieve.
110
C m
F sus
B♭2

114
That's why I be - lieve,
F/G
Gm7
114
Cm
that's why I be - lieve,
Fsus
B♭2
B♭2/A♭
(6)
be - lieve!
B♭2/E♭
Fsus/C
F
B♭

# Optional Ending

## Scene 13

*(After* THAT'S WHAT I BELIEVE *ends,* JOSEPH *walks to the end of the stage. He talks to the audience as if they are a group of people that have gathered to hear about Jesus.* NICODEMUS, PETER, *and other followers are in the background–supporters of the same cause; new Christians united in Christ.)*

JOSEPH: Forgive me friends. I'm not the most articulate man. But what I have to say to you could change your life forever. At one time in my life, I followed all the rules. I upheld the law. I was interested in the power of men. These things weren't bad in and of themselves. But when I met Jesus and He offered something new, I was confused. Should I stay the course? Follow after a dear friend? Maintain a life-long position of authority and security?

Well, I made my choice. I follow Jesus now. And I'm here to tell you that Jesus offers every one of us a new life. A life that isn't void of law, but one that is liberated by love. Jesus gave His very life that we could have this love. He was crucified and buried. I placed Him in the tomb myself. But three days later, He conquered death. He came back to life–the tomb empty–in the grave no more.

I know, friends, that this is hard to believe, but hear me. Jesus loves every one of you, and He died to pay for your sins. His death and resurrection mark the beginning of a new era. A new beginning is available to each of you. All you need to do is believe. Believe that Jesus is God's Son. Don't make the mistake of trying to define on your terms what "messiah" means. Claim Jesus and you'll live free! No matter what your life looks like now, He has something richer and fuller than you could ever imagine.

Jesus was not just a man. He is the Messiah you long for.

*(After* JOSEPH'S *monologue, sing a reprise of* MORE THAN JUST A MAN PART 1, *beginning at measure 44. If you are using the accompaniment track, start at index point 81. When the song ends you may want to have your pastor offer an invitation, give some closing comments and close in prayer. If offering an invitation, consider using* UNDERSCORE 7 (NEAR THE CROSS) *or play a song of your choosing.)*

# GENERAL PRODUCTION NOTES

*More Than Just a Man* is a dramatic musical that uses the choir in various roles throughout the production. Biblical costume for the choir is recommended as the choir serves as townspeople in Scenes 1, 2, 5, 7 and 9.

Most of the cast requires basic biblical dress. However, particular attention needs to be paid to the costumes for each member of the Sanhedrin. CAIAPHAS' costume should show that he is the high priest.

The play requires multiple locales, so design your set for flexibility. Don't try to create in full detail each scene. Since the temple and the street outside are used for several scenes, you may want to make that a central part of the set. The meeting quarters for the council, the crucifixion scene, and other scenes can be suggested with a few props or pieces of furniture. Use lighting to focus the audience on whatever part of stage you are using at a given time.

Like with any play, casting is key to the success of your production. The roles of JOSEPH OF ARIMATHEA, CAIAPHAS, and NICODEMUS are of prime importance. A complete cast list is in the front of this book.

One of the most moving scenes in the play is the wrapping of JESUS' body performed by JOSEPH and NICODEMUS in Scene 11. Be sure to rehearse this scene well. Movements should be natural as if they are happening for the first time, but you will need to choreograph everything to ensure the timing works out. Also, be sure to take the obvious care needed when wrapping the face of the actor playing JESUS. Use a cloth that "breathes" well like linen or gauze.

An optional ending has been provided that offers an additional scene and suggestion for a reprise. This scene provides an evangelistic appeal and allows for your pastor to move right into an invitation and prayer time. You'll find that on page 133 of this book.

Extensive production notes, including cast descriptions, synopses and staging suggestions for each scene, notes on props, costumes, and technical requirements and much more can be found in the *More Than Just a Man Director's Edition* (ME-62A).

MORE *than just a man*

A DRAMATIC EASTER MUSICAL

MORE *than just a man*

A DRAMATIC EASTER MUSICAL

MORE *than just a man*

A DRAMATIC EASTER MUSICAL

MORE *than just a man*

A DRAMATIC EASTER MUSICAL

MORE *than just a man*

A DRAMATIC EASTER MUSICAL

MORE *than just a man*

A DRAMATIC EASTER MUSICAL

MORE *than just a man*

A DRAMATIC EASTER MUSICAL

MORE *than just a man*

A DRAMATIC EASTER MUSICAL

MORE *than just a man*

A DRAMATIC EASTER MUSICAL

MORE *than just a man*

A DRAMATIC EASTER MUSICAL

MORE *than just a man*

A DRAMATIC EASTER MUSICAL

# An Overview of China's Sci-tech Innovation over the Past Decade

**Published by**
**ACA Publishing Ltd.**
**University House**
**11-13 Lower Grosvenor Place,**
**London SW1W 0EX, UK**
**Tel: +44 (0)20 7834 7676**
**Fax: +44 (0)20 7973 0076**
**E-mail: info@alaincharlesasia.com**
**Web:www.alaincharlesasia.com**
**Beijing Office**
**Tel: +86(0)10 8472 1250**
**Fax: +86(0)10 5885 0639**
**Authors:**
**Gu Ruizhen, Huang Xiaoxi et al**
**Editors:**
**David Lammie, Martin Savery and Zhao Daxin**
**Publisher:**
**ACA Publishing in association with China National Publications Import and Export (Group) Corporation (CNPIEC) and Beijing T-Win Consulting Co Ltd**


Published by ACA Publishing Ltd in April 2015



ISBN 978-1-910760-00-0

A catalogue record for An Overview of China's Sci-Tech Innovation over the Past Decade is available from the National Bibliographic Service of the British Library.

# Contents

## Chapter 5 - Transportation Technology



## Chapter 6 - Energy and Environmental Protection



## Chapter 7 - Agricultural Science and Technology



## Chapter 8 - The Earth and the Ocean

# Foreword

Since the 16th National Congress of the CPC, with steady and continued economic and social development, China's science and technology sectors have enjoyed a 10-year period of thriving development and fruitful achievements. Over the past decade, China has continuously increased its input into science and technology, implemented plans smoothly, strengthened fundamental research, and developed its high-tech industries rapidly. All these efforts have enabled China to emerge as a science and technology power in the world. In these 10 years, China scored remarkable achievements in key innovation areas, including manned space flight, lunar exploration, supercomputers, a new generation of high-speed train and super-hybrid rice. These achievements represent a focus on independent innovation and building an innovation-oriented country. We compile and publish this book to demonstrate the latest Chinese achievements in science and technology over the decade and to celebrate the convening of the 18th CPC National Congress.

This book is intended to help readers learn in an all-round way about China's science and technology developments in the past 10 years, particularly cutting-edge achievements in the areas of aerospace, electronic information, biomedicine, advanced manufacturing and new materials, traffic technology, energy and environmental protection, agricultural technology, earth and marine science. It not only details the achievements and explains technical terms, but also records the important scientific and technological innovation events over the past decade. With rich content and high reference value, this is a popular science title that aims to be authoritative, informative and interesting.

To enrich the book and make its content authoritative, we compiled the book based on *China's Science and Technology Development Report*, the exhibition of national major science and technology achievements in the $11^{th}$ five-year plan, and reports in www.xinhuanet.com, CSTNET, *Guangming Daily, Science and Technology Daily, Beijing Daily, People's Posts and Telecommunciations News* and *Tianfu Morning Paper*. We hereby express our thanks to the above-mentioned media and websites.

This book collects or uses for reference the articles and research findings of the following authors: Gu Ruizhen, Huang Xiaoxi, Li Yun, Tian Zhaoyun, Wang Jingguo, Zhao Wei, Wang Yushan, Bai Ruixue, Gao Hui, Zhang Xudong, Chen Yuming, He Zongyu, Wang Min, Cai Jinman, Yu Xiaojie, Cheng Zhuo, Ren Ke, Chen Weiwei, Xu Zuhua, Zhang Jiansong, Shang Chunya, Zhang Xiaoqi, Jiang Guocheng, Zhang Xinxin, Liu Juhua, Wang Chuanzhen, Zhao Ruixi, Hong Liming, Chen Hao, Huo Peng, Wang Xiaolei, Zhang Qin, Mou Xu, Liu Long, Ming Xing, Wu Jingjing, Wu Jun, Hu Hao, Luo Sha, Qin Qiu, Zhang Jianhua, Zhang Duo, Yu Wenjing, Yuan Junbao, Gao Bo, Cai Min, Tang Ting, Min Guoquan, Li Ping, Qiu Yi, Yu Jingjing, Zhu Liyi, Dong Changqing, Li Junyi, Li He, Hai Mingwei, Zheng Xiaoyi, Wang Pan, Zhou Mian, Tan Jian, Li Yan, Dong Jun, Zhao Wanwei, Lin Ying, Zhan Yuan, An Bei, Li Jizhi, and Hu Junchao (listed in random order). We hereby express our thanks to them.

Please kindly oblige us with your comments.

Compilers

30 June 2012

# Opening remarks: China bravely advances in technological innovation

At 10:03am on 29 June 2012, the Shenzhou IX spacecraft brought the Chinese nation's astronomical dream onto a new stage by landing smoothly onto the grassland at the main landing site in Siziwang Banner county, Inner Mongolia, after 303 hours and 16 minutes of space travel.

China had achieved success in its first manned space docking. The vast outer space engraved a nation's glory and dream of flying into space once again.

On 27 June 2012, China's manned deep-sea submersible, the Jiaolong, successfully reached a depth of 7,062 metres in the Mariana Trench, a world record for manned submersibles.

From the heavens to the deep sea, China has never stopped on the path of innovation. It has scaled the heights of science and technology and explored the mysteries of the ocean depths.

In just 20 years, China successfully made breakthroughs and achieved the three basic technologies of manned space flight: manned ground-space transportation, extra-vehicular activity and docking technology, and has become the third country in the world to master docking technology independently.

Zhang Bonan, chief designer of the Tiangong-1 space module and the manned spacecraft system at China Aerospace Science and Technology Corporation, says China's manned space flight has a pragmatic development plan with clear objectives. From Shenzhou I to Shenzhou IX, every manned spacecraft was set specific goals and tasks, all aimed at the development of a space station. During this process, he says, all the key technologies were grasped independently.

The launch model of "Three Verticals and One Long-range" (vertical assembly, vertical checkout, vertical transportation, long-range rocket launch), the internationally-advanced space instrumentation and command network based on IP technology, the space docking structure, the unique space medical engineering system and the Shenzhou spacecrafts were all made and created in China, creating a glorious chapter in the history of space exploration.

Reaching a record depth of 7,000 metres, the Jiaolong manned submersible also reflects independent innovation. This "Chinese Dragon" was designed and built independently by China's engineering technicians. The assembly, alignment and sea trials were also completed by China independently.

To serve the economic and social development of China, science and technology plays an important role.

The Shenzhou IX spacecraft and Tiangong-1 module successfully achieved manned space docking. "Sunway Blue Light MPP", the supercomputer that is based on an independently-developed central processing unit (CPU) and can perform around 1,000 trillion calculations a second, plays a key role in national defence and economic development. The successful development of the deep-water drilling project reaching 3,000 metres deep has transformed China's deep-water oil and gas exploration and development work. China's broadband mobile communication techniques rank prominently in global terms, with its 4G standard becoming internationally recognised. The bioengineering and pharmaceutical industries are growing rapidly, helping enterprises improve their innovation capacity. With the etching machine as an example, the industrialisation of high-end integrated circuit manufacturing equipment has successfully entered the international market. Large high-end manufacturing equipment has strongly supported industrial development in the aerospace, vessel, vehicle, nuclear power and energy sectors. The technological development of the electric vehicle has yielded important results... One after another, these major scientific and technological achievements demonstrate China's independent innovation and its strategic goal of building an innovation-oriented country.

Zhang Renhe, an academic from the Chinese Academy of Sciences (CAS) and researcher at its Institute of Acoustics, said, "Whether it be it a manned space flight or manned deep-diving, both are major scientific and technological projects that demonstrate China's scientific and technological development level." He said that China has always relied on its own innovation.

Besides breakthroughs in the heavens and in the seas, encouraging news has also emerged in other areas, such as an experimental fast reactor, a supercomputer and super hybrid rice. In 2006, in order to promote international scientific and technological competition, and upgrade its industrial structure, China arranged 16 major national science and technology projects. These covered key areas including electronics, information, biology, medicine, energy and environmental protection.

Every high-tech advance injects new vitality. Some 80% of China's 1,100 new materials were developed under the guidance of space technology. Almost 2,000 space technology achievements have been applied in areas such as satellite navigation and communication, meteorology, disaster prevention and mitigation, and food production.

"When the first satellite entered space, no one could predict that the GPS positioning system would develop into a huge industry," commented Zhang Renhe. He also said that Jiaolong's voice communication and audio and video transmission system has been adapted for use in underwater communication in the Qingdao Olympic Regatta.

The substantial increase in scientific and technological strength makes China a scientific and technological power with significant influence. China now leads the world in terms of R&D personnel, the number of papers published internationally, and number of authorised patents. Major innovation results such as super hybrid rice and the supercomputer show intense vitality, demonstrating China's bright outlook in scientific, technological, economic and social development.

# Chapter 1

# Aerospace

In the past 10 years, China has rapidly developed its aerospace industry: manned space flights, lunar exploration and other major aerospace technology projects have made dramatic breakthroughs. The overall space technology level has risen remarkably; the economic and social benefits resulting from space application have been significantly improved; space science has achieved innovative results.

China's aerospace sector has developed rapidly, from the first space travel in 2003, to the first extra-vehicular activity in 2008, to the first successful automatic docking between Tiangong-1 and Shenzhou IX in 2011, and then to the successful docking under manual control between Tiangong-1 and Shenzhou VIII in 2012. During the 12th five-year-plan, China will complete around 20 space launches annually. In 2012 alone, 21 rockets and 30 satellites were launched.

Manned space flights have spawned numerous benefits. The lunar exploration programme in particular has generated positive headlines.

On 24 October 2007, China successfully launched its first lunar probe, Chang'e-1, which achieved the goal of "precisely changing orbit, successfully circling the moon", acquiring scientific data and the entire map of the moon's surface, and successfully implementing a "controlled crash" onto the moon. On 1 October 2010, China successfully launched the Chang'e-2 lunar probe, which obtained a full map of the moon's surface with high resolution and high-definition images of the Sinus Iridum area, and successfully completed several extended experiments such as circling one of the Lagrange points-L2. These achievements lay a foundation for the implementation of more deep-space exploration missions.

Moreover, earth observation satellite data have been widely used in areas of economic and social development; the application of telecommunication satellites is making steady progress; BeiDou Navigation Satellite System, China's GPS, now covers China and its neighbouring region.

## Manned space fight

### *Interpretation of key terms*

The Project 921 manned space flight project was approved by the CPC Central Committee, the State Council and the Central Military Commission in 1992. It was the largest, most complicated and technically difficult project in China's aerospace history. The first step in this project was to launch a manned spacecraft, to build up an experimental manned spacecraft project and to conduct space application experiments. The second step was to achieve breakthroughs in manned spacecraft and docking between spacecraft after the successful launch of the first manned spacecraft, to use manned spacecraft technology to build a space lab and send it into space. The third step was to build a space station and maintain it over the long term.

The manned space flight project consisted of eight systems: astronauts, space applications, a manned spacecraft, carrier rocket, launch site, communication monitoring, landing site and a space lab. Across the country, there are 110 strong research institutes conducting research and construction work, and more than 3,000 units providing support tasks.

The project's research and construction work started in 1992. In November 1999, the first unmanned flight experiment was conducted. Subsequent unmanned flight experiments were conducted in January 2001, March 2002 and December 2002.

On 15 October 2003, astronaut Yang Liwei boarded the Shenzhou V manned spaceship and completed China's first manned space flight.

From 12-17 October 2005, astronauts Fei Junlong and Nie Haisheng boarded the "Shenzhou VI" manned spaceship and completed China's second manned space flight.

From 25-28 September 2008, astronauts Zhai Zhigang, Liu Boming and Jing Haipeng boarded the Shenzhou VII manned spaceship and completed the first extra vehicular activity by Chinese astronauts, making China the world's third country to master the key technology required for extra vehicular space activities.

In September and November 2011, the Tiangong-1 space lab module and the Shenzhou VIII spaceship successfully conducted China's first space rendezvous and docking experiment.

At 6.37pm on 16 June 2012, the Shenzhou IX spaceship lifted off smoothly at Jiuquan Satellite Launch Centre. On 24 June, in the near-earth orbit 340km from earth, three Chinese astronauts – Jing Haipeng, Liu Wang and Liu Yang – completed the manual-controlled docking between Shenzhou IX and Tiangong-1. This showed China had mastered docking technology, the last of the three basic manned space flight technologies after manned ground-space transportation and extra-vehicular activity.

**Future planning**

By 2016, research, development and a space lab launch; mastering key space station technologies such as astronauts' medium-term stay; Conduct a certain scale of space applications.

Around 2020, research and development will achieve a launch core module and experiment module; the assembly of a 60-tonne manned space station in orbit; a breakthrough and mastering of the building and operation technology of a near-earth space station, as well as the long-term manned flight technology in near-earth space; carrying out large-scale space applications.

## *Looking back at the highlights*

### From Shenzhou I to Shenzhou IX: nine launches, nine breakthroughs

***Shenzhou I: sample***

Launch time: 06:30:07, 20 November 1999

After all on-ground interference calculations and simulation tests have been completed, China needs to send a spacecraft into space to verify its previous achievements, and obtain more test data that can't be achieved on the ground, such as heat-resisting technology and recycling technology.

The mission carried out by Shenzhou I was historic. As China's first independently-developed spacecraft, Shenzhou I, eight meters long and weighing 7,755kg, successfully completed its historic mission: to assess the carrier rocket's performance and reliability, verify the accuracy of the spacecraft's key technologies and systems design, and coordinate the whole system's functions including launching, telemetry and telecontrol communication, recovery and landing.

At 03:41 on 21 November 1999, when the re-entry module of Shenzhou I landed on Inner Mongolia's grasslands, the Chinese manned space flight era began.

***Shenzhou II: shaping up***

Launch time: 01:00, 10 January 2001

The structure, technical performance and requirement of Shenzhou II are similar to that of a manned spacecraft. All adapted equipment relevant to an astronauts' life-support are real. Therefore, Shenzhou II was known as the "flight model" by Chinese aerospace personnel.

As China's first unmanned spacecraft, Shenzhou II consisted of an orbital module, a re-entry module and a propelling module, with a newly-extended system structure and enhanced technical functions.

Shenzhou II carried out experimental duties in many areas. For the first time, China conducted experiments in a micro-gravity environment inside the spacecraft in the areas of space life science, space materials, space astronomy and physics, including experiments on space environment effects on plants, animals, aquatic organisms, microorganisms, along with cell lines and cellular tissue.

***Shenzhou III: "manned"***

Launch time: 22:15, 25 March 2002

Shenzhou III did not carry a real person, but a dummy that could simulate the breathing of an astronaut and many other important physiological activity parameters when living in space, such as cardiac rhythm, blood pressure, oxygen consumption and heat emission in the blood circulation system. China's first "anthropomorphic dummy" had a similar physique to real astronauts, and was equipped with a human metabolism simulation device and a physiological signal device. After the rocket was launched and the spacecraft entered its orbit, the dummy sent back various indices, providing valuable data to support astronauts entering outer space.

Shenzhou III was equipped with escape and emergency life support functions for astronauts, and had an improved parachute system, so as to ensure a safe landing on return.

Shenzhou III carried 44 payload equipment items in 10 categories. After completing several scientific experiments, it went back to the main landing site at 04:51 on 1 April 2002, and achieved full success.

***Shenzhou-IV: low temperature***

Launch time: 00:40, 30 December 2002

That winter, late at night, the ground surface temperature at Jiuquan Satellite Launch Centre was as low as -29°C and boilers were operating at full power. The staff on the launch pad were exposed to the freezing outdoors. But the Shenzhou-IV coped with the low temperature and broke China's record for a low-temperature launch.

Shenzhou-IV had a similar layout, functions and technical conditions as the three previous launches, and it solved similar problems, such as excessive toxic gas and space radiation. The experiments included a space application system for manned space flight, an astronaut system, spacecraft environment control and a life support sub-system. Several research programmes were also completed. That is to say, if Shenzhou-IV was launched successfully and returned as planned, the time for Chinese people to travel in space would be approaching.

In fact, this is what happened.

***Shenzhou V: the dream comes true***

Launch time: 09:00, 15 October 2003

Although it stayed in outer space for only one day and with only one astronaut on board, Yang Liwei, Shenzhou V realised a long-held dream of the Chinese people. At around 7am on 16 October 2003, Yang Liwei, China's first astronaut, emerged from Shenzhou V's re-entry module, and waved his hands to the waiting crowd in Inner Mongolia.

Zhao Jianwei, Xinhua News Agency

**At 18:40 on 15 October 2003, Chinese astronaut Yang Liwei sends his regards to the people of the world from outer space, and displays the national flag of the PRC and the flag of the United Nations**

For Shenzhou V, the "one-day, one-person" flight represented a comprehensive assessment of its working performance, reliability, safety and inter-system coordination systems. Chinese astronauts achieved excellent results in the assessment.

China became the world's third country to carry out a manned space flight, after the Soviet Union and the US. Before conducting a manned space flight, America experimented eight times; Russia five times and China four times.

***Shenzhou VI: somersaults***

Launch time: 09:00, 12 October 2005

Astronaut Fei Junlong took about three minutes to complete four somersaults in outer space. With Shenzhou VI travelling at 7.8km per second, each of his somersaults was equivalent to a distance of about 351km on earth.

Shenzhou VI's manned space flight involved two-astronauts and took longer than one day. During this space flight of nearly five days and five nights, the two astronauts conducted a series of space science experiments, including passing through the orbital and re-entry modules, ergonomic evaluation, medical experiments and spaceship equipment manoeuvring in the orbital module.

***Shenzhou VII: performs EVA***

Launch time: 21:10:04, 25 September 2008

Forty-three hours after the launch, Astronaut Zhai Zhigang began to conduct

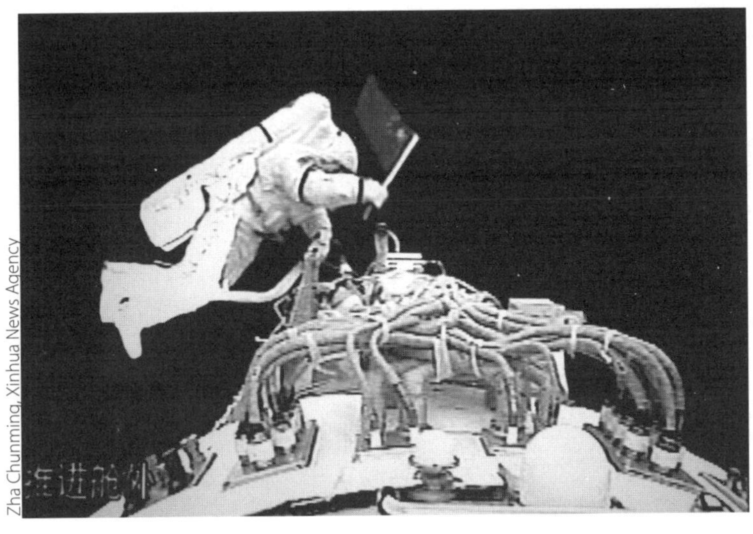

Zha Chunming, Xinhua News Agency

Astronaut Zhai Zhigang, who executed the task of Shenzhou VII's extra-vehicular activity, waves the national flag of China after going out of the module

China's first extra vehicular activity (EVA). It is estimated that nearly 1bn people around the world watched his step into space. For a long time after, many people repeated the sentence he uttered in outer space: "I have been out of the module. I feel good."

Shenzhou VII's manned space flight completed extra vehicular activity with astronauts and an accompanying mini-satellite, and completed several technological experiments successfully. The extra vehicular activity, lasting 19 minutes and 35 seconds, made China the third country in the world to grasp this technology.

***Shenzhou VIII: rendezvous and docking***

Launch time: 05:58 am, 1 November 2011

The major mission of Shenzhou VIII was to catch up and "kiss" the Tiangong-1 space lab module. Although it was unmanned, Shenzhou VIII still attracted considerable attention. That's because docking is an important milestone, an essential link in a manned space flight project.

At 01:36 on 3 November, Shenzhou VIII and Tiangong-1 achieved their first space rendezvous and docking. At 20:00 on 14 November, under the watchful eye of Beijing Aerospace Control Centre, Shenzhou VIII and Tiangong-1 successfully completed a second docking. This second manoeuvre verified the functions and performance of docking measuring equipment and the docking mechanism. At 19:32 on 17 November, Shenzhou VIII landed at the main landing site in Siziwang Banner county, Inner Mongolia.

***Shenzhou IX: docking under manual control***

Launch time: 18:37, 16 June 2012

The mission of Shenzhou IX was to transport three astronauts – Jing Haipeng, Liu Wang and Liu Yang – to Tiangong-1, and complete docking under manual control. At 14:14 on 18 June, Shenzhou IX and Tiangong-1 successfully achieved automatic docking. At 17:07, the astronauts boarded the Tiangong-1 experiment module, the first time a Chinese astronaut had visited an on-orbit aircraft. On 24 June, in the near-earth orbit more than 340km from earth, three Chinese astronauts completed a manual-controlled docking between Shenzhou IX and Tiangong-1. At 10:03 on 29 June, after 303 hours and 16-minutes of travel in space, the Shenzhou IX spaceship landed at the main landing site in Siziwang Banner county, Inner Mongolia. China's had successfully achieved its first manned space docking.

This achievement showed that the Chinese had acquired the ability to send people and logistics supplies to on-orbit spacecraft. China had now achieved the three basic technologies of manned space flight: manned ground-space transportation, extravehicular activity and docking technology.

**Highlights of China's first manually-controlled space docking**

At around 11:00 on 24 June 2012, Jing Haipeng, Liu Wang and Liu Yang, all wearing milky white space suits, made their final preparations ahead of a manual-controlled rendezvous and docking inside the re-entry module. They were each holding a flight manual. Liu Yang carefully tucked her hair behind her ears.

Six days earlier, after Shenzhou IX and Tiangong-1 completed an automatic rendezvous and docking. They conducted aircraft manipulation, life care and some medical experiment programmes, acquiring lots of valuable data.

Although the technology of a spacecraft's automatic space docking was realised in the Shenzhou VIII flight, that didn't mean that China had mastered space docking technology; automatic control and manual control by astronauts are the two basic docking methods used in international manned space flights. They are back-ups for the other method and neither is dispensable.

At this time, developments in exploration were still moving forward. At Beijing Aerospace Control Centre, sunlight poured in and a colourful halo shone on the centre's giant screen like a dream.

Ma Yongping, deputy director of Beijing Aerospace Control Centre, said, "The whole course of the spaceship from the 400-metre mooring point to the docking mechanism contact located in the sunlight zone. This is a more rigorous test for measuring the accuracy and anti-light interference ability of a spaceship's optical equipment in strong sunshine."

The spaceship calmly moved to the next stop, the 140-metre mooring point, and had 45 minutes for rest and reorganisation. The preparation work was going on in an orderly manner.

The spaceship stretched out to the docking mechanism, just like reaching out its "hands" to an old friend, Tiangong-1. This was China's most complicated space mechanism so far.

Liu Wang gently manoeuvred the control handles on both sides of the seat. "Number 1 control handle works normally." "Gesture handle works normally."

Every time a normal state was confirmed, Liu Wang and Liu Yang would show a thumbs up to the camera...

**On 24 June 2012, Shenzhou IX and Tiangong-1 successfully completed their first manually-controlled rendezvous and docking**

Zha Chunming, Xinhua News Agency

At 12:42, the spaceship set out again. At that time, Shenzhou IX had changed its state from automatic control to manual control. That is to say, its every step would be manipulated by the astronauts.

Astronaut Liu Wang's figure appeared clearly on the large LED screen at Beijing Aerospace Control Centre. The spaceship handle seemed to be an extension of his hands.

If it could be said that the Chinese astronauts entering outer space previously were "sitting in the spaceship", Liu Wang was able to "manoeuvre the spaceship". On the day after the traditional Dragon Boat Festival, he was a helmsman rowing a dragon boat.

It's hard to be a "helmsman": Liu had to control the angle between the spaceship and Tiangong-1 to within 1°, and the horizontal deviation no more than 0.2-0.3 metres. For a spacecraft weighing more than 8 tonnes and 9 metres long, the task is similar to "threading a needle", a huge challenge for an astronaut's hand-eye coordination and psychological stability.

Liu Wang, seated in the centre, was manipulating handles while watching instruments. Adjust, forward, adjust. The 1,500-odd ground simulation training sessions helped make him calm and unfazed.

Bai Yanqiang, director of the China Astronaut Research and Training Centre, said, in order to ensure a smooth docking in any condition, Liu was tested with countless emergency situations on the ground, including different distances, different light conditions, different initial deviation. Liu passed each of these difficulties with a 100% success rate.

As the youngest "little brother" among China's first batch of astronauts, 43-year-old Liu Wang made his dream come true through his superb skills, and shouldered the task of a manually-controlled rendezvous and docking, after a 14-year wait.

The Shenzhou IX spaceship unfolded its blue wings and flew to Tiangong-1 in a soft light-purple halo. Commander Jing Haipeng, sitting on the right side of Liu Wang, set his eyes on the instrument panels. As China's first astronaut to fly in space twice, he provided support for Liu Wang and the female astronaut Liu Yang, who joined the team for the first time. Closer, closer... 20 metres, 10 metres, 5 metres... the spaceship was fast approaching Tiangong-1.

At 12:48, Shenzhou IX's docking ring linked with Tiangong-1. They looked like two birds in flight, touching each other softly, skimming far above Qinghai province, and continued to fly day by day.

Flying far above Gansu and Shaanxi, just like the Weihe River flowing through the fields of the two provinces, 12 docking locks on the two spacecraft were started accurately. At 12:55, Tiangong-1 and Shenzhou IX held each other's hands closely. The docking succeeded!

Liu Wang, Jing Haipeng and Liu Yang also held hands. Jing Haipeng was calm, as usual. Liu Yang smiled like a summer flower. Liu Wang waved his clenched fist once more, just like the scene eight days previously when they went on an expedition to the Asking Heaven Pavilion deep in the desert. But this time, there was one difference: tears glimmered in his eyes.

Thunderous applause erupted in the Beijing Aerospace Control Centre. Liu Boming, who had been on the space expedition with Jing Haipeng four years previously, couldn't contain his excitement: "A great success!"

Zhou Jianping, chief designer of China's manned space flight project, said, "After making breakthroughs in manned ground-space transportation and extravehicular activity, China has mastered docking technology, the last one of the three basic technologies of a manned space flight. It means that we've got the basic ability to build a space station, moving another steady step forward towards the goal of the three-step development strategy for manned space flight".

## Knowledge link

### Behind the precise rendezvous and docking

"The astronauts completed the rendezvous and docking between the spaceship and Tiangong-1 through precise manual control. This is 'manipulating' a spaceship in its true sense," said Chen Jianmin, a measurement and communication expert from China Electronics Technology Group Corporation (CETC). Astronauts that conduct an operation rely upon a large quantity of measurement and communication system equipment.

Spatial precision: deviation of no more than 1mm

Chen Jianmin said, "To make a rendezvous and docking succeed, the orbit should be determined in high precision through space-ground remote metering, and the docking position of Tiagong-1'and Shenzhou-IX should be confirmed and aligned, with a deviation of no more than 1mm. This requires a lot of ground observation stations."

In outer space, Tiagong-1 and Shenzhou-IX operated at high speed, travelling at more than 28,000kph. During the docking process, even a tiny error could make the spaceship deviate from the target spacecraft, or cause a rear-end collision. Chen Jianmin said, "Ground stations continuously send and receive electric waves, measuring a spacecraft's position by calculating the spacecraft-earth distance and electric wave transmission speed. After the successful launch of Shenzhou VII, we had a makeover of measurement and communication equipment at ground observation stations. Now the equipment has a higher speed and measuring accuracy."

Time accuracy: no more than 1 second in 3,170 years

"For the measurement and communication system, time is very strict. The first cosmic velocity is 7.8km per second. Spaceships at this altitude should keep a minimum speed of 8km per second. The higher the orbit, the higher the speed". If timing is not accurate, large errors occur. A miss is as good as a mile," said Chen Jianmin.

Synchronising the time for an entire measurement and control network is equivalent to setting the clock in our daily life. "How high is the clock's precision? It's 10 to the power of 11, approximately no more than 1 second in 3,170 years," said Chen Jianmin.

Space-ground communication: "country footpath" becomes "expressway"

The television images seen and the voices heard from the ground by astronauts during a manually-controlled rendezvous and docking process are largely due to the communication systems being unobstructed. Spacecraft, ground measurement and control stations, and the command and control centre set up an invisible space-ground communication network. Through this network, a smooth information transmission can be realised including various communication and measurement data, command and dispatch, voice communication and video.

"In the past, astronauts couldn't see ground images; now, space-ground and station-station communication systems on the ground have realised networked communication, entering a new interconnection and interworking age," said Yang Zhiguo, a senior engineer for China Electronics Technology Group Corporation (CETC). He added, "If in the past, communication was likened to a 'country footpath', current communication is a two-way 'expressway', which can complete space-ground information transmission in larger quantity and quicker. Information transmission quality is also of a higher frequency and is greatly advanced. Now, images transmitted can compare favourably with HD TV images, with the voice clarity better than everyday telephone connection quality."

During the docking process between Shenzhou IX and Tiangong-1, the data in each second is important, requiring a completely reliable transmission network.

> "When transmitting data via the IP network for civil use, there would be a 'packet loss' phenomenon, i.e. 'data loss'. This kind of thing can't be allowed to happen after the spaceship is launched, because it would cause error in relation to the directives or connection interruption. Therefore, the 'packet loss rate' must be zero," said Lu Huabin, senior engineer for CETC. During the launch of Shenzhou IX, both the network equipment and circuits have back-ups, which mean two networks operating at the same time, ensuring absolute data transmission safety.

## Daily highlights: replay of scenes in Tiangong-1's seven days

From the "checking in" of Tiangong-1 on 18 June 2012 to going back there after the manual-controlled rendezvous and docking on 24 June, the three astronauts experienced a complete cycle from Monday to Sunday.

Seven days in Tiangong-1: there were different highlights each day.

### Day 1 - Astronauts enter Tiangong-1

[Camera Scene] The vast space is like a boundless ocean. At 17:00 on 18 June, Jing Haipeng, Liu Wang and Liu Yang "swam" into Tiangong-1.

The first "visitor" Jing Haipeng waved at the camera, and then gently pulled Liu Wang behind him. Light as a feather, Liu Yang turned round and faced the camera with the help of her two team-mates. The long-term joint training made every cooperative movement between them so natural.

Wave hands, and smile. 343km from earth and above Xiamen Measurement and Control Station, the three astronauts posed for a "space family portrait".

A new page in China's manned space flight was unveiled. The Tiangong, which means Heaven Palace, had delivered what the Chinese people had been waiting thousands of years for.

[Interpretation] Wu Ping, press spokesman for China's manned space flight project, said, "The task this time is to let the astronauts visit an in-orbit spacecraft for the first time, and complete the transportation and replenishment of personnel and goods from the ground to an in-orbit spacecraft. Meanwhile, the ability of Tiangong-1 to support the astronauts' work will also be assessed. What will also be examined are the functions and performance of the environmental control and stable operation.

### Day 2 - Tiangong-1 receives first email from the motherland

[Camera Scene] At 15:46 on 19 June, Shenzhou IX astronauts successfully received their first email from the motherland. The first space-ground email transmission experiment was a success.

Beijing Aerospace Control Centre sent a carefully-selected data package, including pictures, documents and videos to Tiangong-1 via a special software and hardware platform. Later, the astronauts saw their first email from the ground in the terminal display system.

That day, the astronauts received five emails, including various pictures and a 10-second-long animal video.

[Interpretation] Li Jian, deputy chief engineer for Beijing Aerospace Control Centre, said, "Compared with sending a remote control directive, an email is more convenient and more flexible. In this way, the astronauts can correspond with family members, and also receive the movies and music that they like. They can also receive instructive video and a flying plan sent from the ground."

**Day 3 - Astronauts use "space bicycle" for the first time**

[Camera Screen] At noon on 20 June, Liu Yang changed from a blue work uniform to a sky-blue, short-sleeved sweater and dark-blue shorts, looking refreshed and neat.

With help from Liu Wang and Jing Haipeng, Liu began to ride a "space bicycle", which was a tailor-made exercise machine for astronauts. When riding the bicycle, Liu Yang kept a steady motion, turning her head to the camera at times and waving her hands.

This was the first time Chinese astronauts have used a bicycle ergometer.

[Interpretation] Li Yinghui, deputy chief designer of the astronauts' system, said, "Compared with former Chinese manned space flights, this time, the astronauts' dwelling time in space increased from several days to more than 10 days. Space microgravity became the main danger. The "space bicycle" and other exercise facilities can help astronauts counter the effects of space microgravity.

**Day 4 - Blessing of love from space**

[Camera Screen] On the morning of 21 June, Jing Haipeng and Liu Wang got up very early to take over duty from Liu Yang, who was on her first night shift. Several hours later, Liu Yang woke from sleep and carefully wiped her face and hair with a towel.

According to a space-ground synchronised work and rest regime, the three astronauts worked from 8am to 8pm every day. They arranged duty shifts, three meals a day and took turns on duty. Each day, they worked for about eight hours, slept for about eight hours, performed daily care for about six hours, and relaxed for about two hours.

At dusk, the three astronauts made video calls to their families, with eight minutes for each family. Although the calls were private, it can be seen from the images that everybody was quite happy. It was the birthday of Liu Wang's wife that day. Liu Wang, a music fan, played the harmonica to send his birthday message.

[Interpretation] Liu Weibo, director of overall research office for astronauts' system: "Tiangong-1 orbited the earth every one-and-a-half hours. Therefore, the astronauts experienced a day and a night every 90 minutes. Arranging tasks according to the space-ground synchronisation principle in space was favourable to the astronauts' health. They worked in the daytime, and rested at night."

**Day 5 - Astronauts "drive" a spaceship for the first time**

[Camera Screen] On 22 June, under the normal flying state of the conjoined Tiangong-1 and Shenzhou-IX, Liu Wang manipulated the gesture handle and conducted control test for the assembly's flying state.

At around midnight, Liu Wang manipulated the gesture handle and implemented three kinds of control gestures: yaw, roll and tilt. Each implementation got a precise reaction from the assembly. Meanwhile, inside the re-entry module, Jing Haipeng was supporting Liu Wang's manual-controlled operation. Liu was monitoring the test process in the experiment module of the Tiangong-1.

This was the first spacecraft control gesture carried out by Chinese astronauts.

[Interpretation] Chen Shanguang, general director of the astronauts' system, said, "Spaceship-driving" needs close coordination between astronauts. During their days in space, the astronauts in Shenzhou IX had good teamwork, accomplishing tasks excellently, including spacecraft operation, life care and space medical experiments, and gained a large quantity of valuable data through the experiments. The three astronauts' bodies and psychology were all in very good condition."

**Day 6 - Astronauts send Dragon Boat festival message**

[Camera Screen] "Happy Dragon Boat Festival". When Jing Haipeng manually controlled the flight with these words facing the module's camera on Tiangong-1, applause broke out at Beijing Aerospace Control Centre.

23 June was the fifth day of the fifth month in the Chinese lunar calendar. At around midday, taking a break from scientific experiments, the three astronauts gathered in front of the Tiangong-1 camera, and called the astronaut centre's ground staff to send festival greetings. On behalf of the three astronauts, Jing Haipeng sent greetings to the Chinese people. "On the traditional Chinese Dragon Boat Festival, we the three astronauts wish Chinese in and outside of China a happy holiday!"

This was the first time that Chinese astronauts had spent the traditional Chinese festival in space.

[Interpretation] Jia Wenjun, head of space-ground communication system at the Beijing Aerospace Control Centre, said, "From Shenzhou IX, images can be sent from ground to space, a breakthrough achievement. The strengthening of transmission ability means a lot for the astronauts' space life. They can see images from the ground, and have face-to-face communication with family members and ground support personnel. For future astronauts staying a long time in orbit, it is a great intellectual stimulus and psychological support.

**Day 7 - The success of manual-controlled rendezvous and docking**

[Scene] In the ink-coloured space, Tiangong-1 and Shenzhou IX were eye-catching, like two white rocks flying high.

Closer, closer... At 12:55 on 24 June, with another body contact between Tiangong-1 and Shenzhou IX manoeuvred by the three astronauts, China's first manual-controlled rendezvous and docking was achieved.

The hands of Liu Wang, Liu Yang and Jing Haipeng were tightly held together for quite some time. "Driver" Liu Wang, sitting in the middle, waved his clenched fist once more, just like the scene eight days before when they went on an expedition in the Asking Heaven Pavilion deep in the desert. But this time, there was one difference: tears glimmered in his eyes.

Thunderous applause rose up at Beijing Aerospace Control Centre. Liu Boming, who went on the space expedition with Jing Haipeng four years previously, couldn't contain his excitement: "A great success!"

Around 14:00, the three astronauts went back to the Tiangong-1 and continued their brilliant stay there.

[Interpretation] Zhou Jianping, chief designer of China's manned space flight project, said, "In terms of the development of the world's manned space flight, a rendezvous and docking should cover two aspects: automatic and manual controls, which are a back-up for each other, neither is dispensable. Only after both automatic and manual technologies are verified, can it be said that a complete rendezvous and docking has been realised. After making breakthroughs in manned ground-space transportation and extravehicular activity, Chinese aerospace researchers and staff have entirely mastered docking technology, the last one of the three basic technologies of manned space flight. It means that China had successfully opened the gate to the space station age."

## Lunar exploration

### *Interpretation of key terms*

In October 2007, China successfully launched the Chang'e-1 satellite, and realised engineering goals and a scientific detection task. After an artificial earth satellite and manned space flight, the full success of the moon exploration project set the third milestone for China's aerospace adventure.

As a landmark project for China's scientific and technological innovation, the moon exploration project is one of the key projects in China's long-and-medium-term scientific and technological development. According to the plan, there will be three periods of unmanned lunar exploration by China before 2020.

The moon exploration project was organised and implemented by SASTIND. It was named the Chang'e Project, taking Chang'e as its code name, and numbered in sequence.

**Project planning**

The lunar exploration project is divided into three periods: "circle-around detecting", "soft landing detecting" and "sample return detecting", which are called "circle, land, return" for short.

Circle-around detecting is the first part of the project. Through various remote sensing detecting means, it performs a general global survey. So far, it has been implemented successfully.

Soft landing is the second part of the project. It will realise a lunar landing by China's detector and carry out an inspection tour and exploration, precise detection in certain parts of the moon. Three launches were made: Chang'e-2, Chang'e-3, and Chang'e-4. Chang'e-2 is a technical guide satellite, having already completed its tasks; Chang'e-3, under development, will land on the moon for the first time; Chang'e-4 is the back-up of Chang'e-3.

Sample-return detection is the third part of the project. It will take a sample from the moon for the first time and then return. It will carry out a laboratory test, analysis and comprehensive study of the moon sample. Two launches are planned, Chang'e-5 and Chang'e-6. Now, the development work starts.

## Knowledge link

**Chronicle of events in China's lunar exploration project**

- In 1991, Chinese aerospace experts said that the lunar exploration project should be carried out.
- In 1998, the State Commission of Science and Technology for National Defence Industry (COSTIND) began to plan and demonstrate the lunar exploration project, and carried out preliminary key scientific and technological projects.
- In November 2000, the State Council Information Office issued a white paper, China's Space Activities, in which "carrying out preliminary research of deep-space exploration focusing on lunar exploration" was listed in recent development goals.
- In April 2003, China National Space Administration announced the launch of preliminary research for the lunar exploration project.
- In January 2004, the State Council officially approved the circling-around detection project.
- On 25 February 2004, the leading group of circling detection project held its first meeting, which passed "the General Development Requirements for the Circling Detection Project". The project was named Chang'e.
- On 29 December 2005, the leading group of the Circling Detecting Project held its third conference, in which it agreed that the project had entered the model development stage.
- On 16 July 2006, the 40-metre antenna in the Ground Application System, Kunming, passed inspection.
- On 20 October 2006, the 50-metre antenna in the Ground Application System, Miyun, passed inspection.
- In August 2007, the circling detection project finished the development of the Chang'e-1 and Long March-3A carrier rocket products, and turned to the launch and implement stage.
- On 24 October 2007, Chang'e-1 satellite was launched into space by the Long March-3A carrier rocket at Xichang Satellite Launch Centre.
- On 26 November 2007, China National Space Administration officially released the first moon surface image sent back by the Chang'e-1 satellite.
- On 24 October 2008, the Chang'e-1 satellite finished various tasks during its one year of in-orbit operation and detection, obtaining 1.37TB of scientific detection data.

- On 1 March 2009, the Chang'e-1 satellite successfully performed a "controlled crash onto the moon", drawing a satisfactory conclusion to Phase I of the moon probe project.
- On 11 January 2010, the CPC Central Committee and the State Council held the National Science and Technology Awards Conference in Beijing. Three results including the "circling detecting project" were awarded the Grand prize in the national technology awards.
- On 18 January 2010, the second working conference of the second period of the lunar exploration project was held in Beijing. Chen Qiufa, the project's general manger, and Wu Weiren, the general designer, gave the "Responsibility Paper Task for Chang'e-2" to the general directors and general designers of the five systems.
- In February 2010, all systems of Chang'e-2 finished product development. The satellite entered the assembly test.
- On 10 July 2010, Chang'e-2 satellite was transported to Xichang Satellite Launch Centre.
- On 23 September 2010, Long March-3C carrier rocket was transported to Xichang Satellite Launch Centre.
- On 1 October 2010, Chang'e-2 satellite was launched into space by the Long March-3C carrier rocket at Xichang Satellite Launch Centre.

## *Focusing on Chang'e-2*

Chang'e-2 is the sister satellite of Chang'e-1, but its CCD camera carried on has a higher definition. Other detection equipment was also improved. The lunar data detector was more detailed. At 18:59:57 on 1 October 2010, Chang'e-2 was successfully launched into space at Xichang Satellite Launch Centre.

### Chang'e-2 successfully circles Lagragian Points-L2

Under precise control, and after 77 days flight since leaving the moon on 9 June 2011, at 23:27 on 25 August, for the first time in the world, Chang'e-2, China's second lunar exploration satellite, entered the orbit of Lagragian Points-L2, which is the gravitationally stable position 1.5m km from earth, starting from the moon's orbit.

The successful implementation of the extendibility experiments by Chang'e-2 realised the great leap of China's aerospace flight from 400,000km to 1.5m km. It realised the main engineering technologies and partial scientific goals of the "Kua Fu" plan that has been demonstrated once: observing the spatial environment and its impact on earth from the Lagragian L1 Point. For researching space weather application and space environment warnings, this has groundbreaking significance.

The successful implementation of the extendibility experiments by Chang'e-2 created several "firsts" for Chinese and international aerospace flights: the first international aerospace activity starting from the lunar orbit to detect the Lagragian Points; the first time China had explored outer space beyond the moon; the first time China designed and controlled the transfer and mission orbits of the Lagragian Points, and realised long-distance measurement, control and communication 1.5m km away. Chang'e-2 successfully circled L2 point, marking a breakthrough of China's ability to innovate in lunar and deep-space exploration. China became the third country/organisation to visit L2 Point after the ESA (European Space Agency) and the US.

**Authority interview: how awesome is the Chang'e-2 whole moon map?**

On the first full moon day in the lunar Year of the Dragon, Chinese scientists sent a lavish gift to global astronomy scholars and enthusiasts: a whole moon map with 7-metre definition and 100% coverage obtained by Chang'e-2.

On 6 February 2012, the State Administration of Science, Technology and Industry for National Defence released this major scientific and technological result of China's lunar exploration project. How awesome is the Chang'e-2 whole moon map?

A Xinhua News Agency journalist talked with researcher Yan Jun, director of the CAS Nation Astronomical Observatory, and chief scientist for the scientific application of China's lunar exploration project.

**Internationally, the highest definition whole moon striograph**

**Journalist:** At what international level is the whole moon map in 7-metre definition obtained by China?

**Yan Jun:** The whole moon map in 7-metre definition has the highest definition among all whole moon maps so far released internationally. It is better than other whole moon digital products in aspects such as spatial resolution, image quality, data consistency and integrity, and inlay precision. It stands at an internationally advanced level.

**High-definition moon map: Apollo 11 traces could be seen**

Journalist: What is the precision of 7 metres?

Yan Jun: Compared with the Chang'e-1 camera, the camera carried by Chang'e-2 is far more advanced. The Chang'e-1 whole moon map has a definition of 120 metres, while that of the Chang'e-2 is 7 metres, 17 times higher. The Chang'e-1 camera could only identify craters and stones larger than 360 metres, but the Chang'e-2 could identify objects of less than 20 metres.

We can even see lunar landing traces left by Apollo 11 on the whole moon map.

**Helping the lunar rover find a safe place to land**

**Journalist:** What is the significance of a high-definition whole moon map for implementing a lunar landing plan?

**Yan Jun:** The high-definition whole moon map has an engineering significance. We could select the landing site for a manned lunar landing, and a soft landing of the lunar rover and the lander according to the comprehensive, detailed and reliable data of the area's landforms and the physiognomy that it provides.

In the moon's Sinus Iridum area, where Chang'e-3 is expected to land, Chang'e-2 obtained images with a definition better than 1.3 metres through lowering satellite altitude and adjusting camera position. This could provide more basic data for Chang'e-3 to select a flat setting point.

**Basic and comprehensive data for scientific moon research**

**Journalist:** What value does it have for scientific lunar research?

**Yan Jun:** A high-definition full map of the moon's surface is the most basic material for scientific lunar research.

A high-definition full map of the moon's surface provides detailed features of the area's landforms and the physiognomy of the near and far sides of the moon, the Antarctic region and the Arctic realm, displays the size, distribution, structure and features of the lunar mare, terra, and various impact craters, and explains space time evolution regularity. It is the base of classifying all-moon terrain divisions and geomorphic units, and provides a scientific basis for studying the relative age of different parts on the moon's surface.

Meanwhile, the distribution map of the whole moon's ring structure and linear structure interpreted by the high-definition whole moon striograph could be used to classify the moon's ground structure and research its structural evolution.

Therefore, it has direct scientific value and important reference value for us to study the moon's substance properties, regionalise the moon's geologic structure, draw the outline of the substance ingredients (rocks, minerals, chemical elements) content and distribution diagram on the moon's surface and analyse the features and the cause of the area's landforms and physiognomy and explore the origin and evolution of the moon.

**Chang'e-2 in good condition, more experiments of extendibility could be arranged**

**Journalist:** How is the current condition of Chang'e 2? Any future plans?

**Yan Jun:** Currently, Chang'e 2 has flown from the moon to the Lagrangian points-L2, carrying space environment detection and engineering technology experiments. It is now in good condition. In future, more extendibility experiments could be arranged according to its condition.

For example, detecting charged particles in the earth's farther magneto tail, observing a possible solar X-ray burst, cosmic gamma-ray burst, so as to deepen knowledge of the sun-earth spatial environment.

Meanwhile, key deep-space exploration technologies will be further verified, such as the measurement, control and communication 1.5m km away, so as to accumulate more experience for follow-up projects.

## Carrier rockets and artificial earth satellites

China's aerospace sector is in a high-density launch period. The goals during the 12th five-year plan are 100 rocket launches, 100 satellite launches and 100 satellites in stable operation in orbit. In the last four to five years, around 20 aerospace launch activities have taken place each year. In 2011, China ranked second in terms of aerospace launches in the world. The first three places were taken by Russia (36 launches), China (19 launches) and the US (18 launches).

### *Carrier rocket*

The carrier rocket is an aerospace transport tool consisting of a multi-stage rocket. It is used to send effective loads into pre-determined orbit, including man-made satellites, manned spaceships, space stations and space probes. The whole rocket consists of a rocket body structure, a propulsion system, a guidance and control system, a self-destruction system and a telemetry and orbit measurement system. Each stage has its own rocket body structure and power plant. Inside the tail stage, there is an instrument capsule. What's been installed here is a guidance and control system, a telemetry and orbit measurement system and a large part of a security system. Interstage sections connect all the different stages of rockets. Effective loads are put above the instrument capsule, with a fairing outside.

China's carrier rockets started in the mid 1960s. Now China can launch all kinds of spacecraft, manned spaceships and moon probes covering low-orbit to high-orbit types as well as of different qualities and uses. It has four series of carrier rockets, including the Long March 1, Long March 2, Long March 3 and Long March 4.

Now, the Long March 5 is under development. According to the plan, it will make its maiden flight and shoulder its first carrying task in 2016. It will also provide transport capacity for China to develop a space station and the third part of the lunar exploration project.

The Long March 5 is a brand new rocket. Currently, the near-earth orbit carrying capacity of our rockets is only nine tonnes. But the capacity of the Long March 5 can reach 25 tonnes, which could guarantee the space station's carrying mission, and offer transport capacity security for the construction of China's future space station and the "third period" of the lunar exploration project.

Another characteristic of the Long March 5 is that it is environmentally friendly. The traditional fuel propellants are mainly nitrogen tetroxide and unsymmetrical dimethylhydrazine. The products of combustion could cause pollution. But the new generation of Long March 5 rockets would adopt liquid oxygen and kerosene, whose products of combustion are mainly water and carbon dioxide. Therefore, there is less pollution.

The Long March 5 rocket is similar to the Ariane-5 rocket, and even has some stronger functions.

## *Earth observation satellites*

Several series of satellites have been developed, such as the Fengyun, HY, ZY, Yaogan and Mapping Satellite, as well as "a constellation of small satellites for environment- and disaster-monitoring and forecasting". Fengyun meteorological satellites have global, 3D and multispectral quantitative observation ability. Fengyun-2 geostationary meteorological satellite carries out double-satellite detecting and back-up for each other. Fengyun-3 is a polar-orbiting meteorological satellite. It realises double-satellites (morning and afternoon satellites) network observation. The imaging width of the HY sea colour satellite is doubled, with a greatly shortened revisiting interval. The first HY dynamic environment satellite launched in August 2011 possesses all-weather and all-time microwave observation ability. The spatial resolution and image quality of ZY satellites are greatly improved. The constellation of small satellites for environment- and disaster-monitoring and forecasting has disaster-monitoring ability in medium definition, wide coverage and high revisiting frequency. In 2010, the major scientific and technological project of the high-definition earth observation system was officially launched.

### The Fengyun clan of meteorological satellites

The Fengyun series of meteorological satellites have applications in many key areas such as weather forecasting, climate prediction, natural disaster and environmental monitoring, resource development and scientific research. They are also extensively used in industries including meteorology, the maritime sector, agriculture, forestry, water conservancy, transportation, aviation and aerospace. In meteorological support for major events such as the Beijing Olympics, China's 60th National Day celebration, Shanghai World Expo, Guangzhou Asian Games, as well as a relief service for natural disasters such as the Wenchuan and Yushu earthquakes and the Zhouqu mudslide, contributions from meteorological satellites cannot go unnoticed.

## Knowledge link

#### The Fengyun clan generational alternation of meteorological satellites

So far, China has successfully launched 12 Fengyun meteorological satellites. Due to the lifespan of satellites, the Fengyun series of meteorological satellites necessitates generational replacement. Currently, there are seven "living" Fengyun meteorological satellites in in-orbit operation.

The first members of the Fengyun Clan were the Fengyun-1 polar-orbiting meteorological satellites. Their main task was getting atmosphere, cloud, land and ocean materials in and outside China, collecting data that were used for weather forecasting, climate prediction, natural disaster, environmental monitoring, etc.

The development of Fengyun-1 did not go smoothly. The Fengyun-1A satellite launched in September 1988 only operated for 39 days in outer space. It met a premature end when it lost its intended position in space. For the same reason, the Fengyun-1B satellite launched in September 1990 only lasted 158 days in orbit.

Nine years later, China launched the Fengyun-1C satellite in May 1999, with a great advancement in its physical fitness and professional proficiency. Because of its excellent performance, it was accepted by the World Meteorological Organisation (WMO) as a member of the global meteorological observation network. The Fengyun-1C satellite operated for eight years and retired "mission accomplished" in 2007. The Fengyun-1D satellite launched in May 2002 inherited the strengths of the Fengyun-1C and further improved it. It is still operating in orbit.

Fengyun-3 is the Fengyun clan's second generation of polar-orbiting meteorological satellites. It carried a dozen detection instruments including a visible light and infra-red scanning radiometer, infrared spectrometer, microwave radiometer, moderate resolution imaging spectro-radiometer (MODIS), TMI, UVO detector, earth radiation budget detector, etc. They can detect in a global, all-weather, multi-spectral, 3D and quantitative manner, acquiring multiple characteristic parameters about the atmosphere, sea and land surfaces. Compared with Fengyun-1, Fengyun-3 has a superior performance.

The Fengyun-3A satellite was launched in May 2008, followed by the Fengyun-3B satellite in November 2010. These "twins" have been accepted into the new generation of international polar-orbiting meteorological satellites network. Their observation data not only serve China's meteorological observation, but are also shared globally.

Fengyun-2 is Fengyun Clan's first generation GOMS (Geostationary Operational Meteorological Satellite), with diversified functions such as up taking visible light, generating infra-red cloud images, water vapour distribution images, low-rate images and other meteorological information, collecting meteorological, oceanic and hydrological data and acquiring space environment monitoring data.

The development of Fengyun-2 encountered many setbacks and frustrations. In April 1994, during the final factory test for the first satellite of Fengyun-2-01, i.e. Fengyun-2 ahead of its launch, an accident happened and the launch was suspended.

Three years later, Fengyun-2A was successfully launched at Xichang Satellite Launch Centre in June 1997; in June 2000, the Fengyun-2B was successfully sent to outer space. These two are experimental satellites, and play an important role in establishing China's business application system with a stable operation.

Based on Fengyun-2A and Fengyun-2B, Fengyun-2C made significant improvements in multi-channel scan radiometer (MCSR), the main detecting instrument, and in satellite reliability. Then it was successfully launched in October 2004, and went into continuous stable operation from January 2005. More than 20 business products were developed based on the application system, and it became popular among customers.

In December 2006, Fengyun-2D was successfully launched, making China's geostationary operational meteorological satellites (GOMS) realise the in-orbit backup and the operation of a double-satellite network. In December 2008, Fengyun-2E was launched into outer space, which replaced the ageing Fengyun-2C and kept the business structure of the "double-satellite detecting and backup" with Fengyun-2D; the ageing Fengyun-2C was put into use for regional observation.

Li Qing, chief designer of Fengyun-2 and engineering designer for Fengyun-4 for the Shanghai Academy of Spaceflight Technology, said, "Since we have had Fengyun satellites, we have been able to observe every typhoon influencing or landing in our country. With the successful launch of the improved Fengyun-2F, our country will be strengthened by the Fengyun satellites' ability to detect and observe land and sky."

## Image data of the ZY-3 cartographic satellite cover more than 4,578 sq km

By 18 May 2012, ZY-3, China's first civil high-definition survey satellite had sent back 499 tracks of effective image data, covering an area of 45,789,000 sq km, in which 9,324,200 sq km are China's territory.

Since the successful launch on 9 January 2012, several technical indicators of the ZY-3 have reached or been superior to those of the same-kind of cartographic satellite abroad, helping China make a big breakthrough in this area. Its image data has been widely applied in many social areas. Data of more than 600,000 sq km are uploaded onto "sky map", the national public service platform on geographic information, and provide online service of imaging; the data are also used on basic surveying and mapping, resources investigation, landing sites and other aspects. In the forest fire that occurred in Yunnan province, the ZY-3 provides immediate HD images for rescue and relief work, playing an emergency security role effectively.

**China launches three seasats**

[Reported by Journalist Channel, www.chinamil.com.cn] At 06:57am on 16 August 2011, China's first dynamic ocean environment satellite HY-2 was successfully launched into space by the Long March-4B carrier rocket at Taiyuan Satellite Launch Centre. This was China's third successful launch.

On 15 May 2002, China's first seasat, HY-1A, was launched into space by a Long March carrier rocket at Taiyuan Satellite Launch Centre, spending 685 days in orbit. On 11 April 2007, the HY-1B satellite was launched into space at Taiyuan. HY-1B was more technically advanced than HY-1A. The HY-1 series uses visible light and infra-red band equipment to generate ground images, mainly detecting sea colour information. They obtained many ocean remote sensing images and data in orbit, and made a great contribution to China's ocean survey work.

On 16 August 2011, China's first dynamic ocean-environment satellite HY-2 was successfully launched into space. It used the microwave remote sensing technique and could monitor all-weather and all-time marine elements such as sea surface wind, ocean currents, sea waves and temperature, providing direct technical support to various areas including marine hazard prevention, maritime transportation, maritime engineering and marine scientific research. The successful launch of the HY-2 marked a great step forward for Chinese seasats. It has opened up a new area for China's ocean observation, and will greatly improve its abilities in marine regulation, the safeguarding of maritime rights, and maritime research.

The HY-2 satellite project created five "firsts": China's first satellite to obtain dynamic ocean-environment data; China's first high-precision spacecraft; China's first microwave remote sensing satellite undertaking quantitative observation; the first satellite-ground laser communication experiment; the first ocean radiative correction and reality testing of microwave remote sensor. (Shang Chunya, Zhang Xiaoqi)

## Knowledge link

### Chronicle of events for China's seasats

- In 1985, preparations started for the project approval of China's first seasat. Since then, China's seasat project entered the stage of basic research and technical readiness.
- In 1993, the State Oceanic Administration (SOA) started project demonstration and assessment for the seasat project.
- In May 1996, the SOA Seasat Leading Group and Seasat General Establishment were set up.
- On 31 January 1997, Seasat General Establishment completed a comprehensive assessment report and preparations for the approval of the ocean colour satellite project.
- On 30 June 1997, the State Commission of Science and Technology for the National Defence Industry (COSTIND) approved the R&D of "ocean colour satellite", approving the project. The satellite was named HY-A.
- In March 1998, SOA Satellite Ocean Application Service was established, in charge of seasat's ground application system building and ground application research.
- In May 1999, the HY-1 satellite ground application system construction was approved by the State Development Planning Commission.
- In September 2000, the National Satellite Ocean Application Service was established after receiving approval from the Office of Central Institutional Organisation Commission, in charge of establishing the HY-1 satellite ground application system.
- In November 2000, the National Satellite Ocean Application Service and the Fifth Institute of China Aerospace Science and Technology Corporation signed the seasat development contract.
- In March 2002, the development of the launchable satellite in the flight model phase was completed.
- At 09:50am on 15 May 2002, China's first seasat, HY-1A, was launched into space by a Long March carrier rocket at Taiyuan Satellite Launch Centre.
- On 27 May 2002, the HY-1A was located in the quasi-sun-synchronous orbit with a predetermined altitude of 798km.
- On 29 May 2002, ground-receiving stations in Beijing and Sanya got the first track of an ocean colour remote sensing image and verified all functions of the satellite and ground application system.
- On 2 September 2002, the in-orbit test and assessment for HY-1A was complete, the satellite test was completed successfully.
- On 18 September 2002, the handover ceremony of HY-1A was held, as well as the signing ceremony of the development contract for HY-1B. The seasat entered business application phase and a normal development period for the seasat's career.
- On 12 December 2002, satellite data from HY-1A were distributed externally.
- In January 2005, COSTIND and the Ministry of Finance officially replied to the development project for HY-1B. In July that year, COSTIND approved the general requirements for HY-1B.

- In March 2007, the HY-1B and Long March-2C carrier rocket left the factory, and were transported to Taiyuan Satellite Launch Centre by a special train. There, the hoisting work was done.
- On 10 April 2007, all preparations were complete before the launch.
- At 11:27am on 11 April 2007, the better equipped HY-1B satellite was launched into space by the Long March-2C carrier rocket at Taiyuan Satellite Launch Centre.
- At 06:57am on 16 August 2011, China's first dynamic ocean-environment satellite, HY-2, was successfully launched into space by the Long March-4B carrier rocket at the Taiyuan Satellite Launch Centre.

### *Communications satellite*

Breakthrough technologies include a large-volume geostationary satellite public platform, space-based data relay, measurement and control. The satellites' technical performance is remarkably advanced. Communication level on voice, data, radio and television are further enhanced. The successful launch and steady operation of the Zhongxing-10 satellite have greatly improved the power and volume of China's communication satellites. The successful launch of the Tianlian-1 data relay satellite gives China the ability of space-based data transmission and space-based measurement and control service to the spacecraft.

### *BeiDou navigation satellite system*

The BeiDou navigation satellite system is a global satellite navigation system independently developed and operated by China, which together with the US's GPS, Russia's Glonass and the EU's Galileo are the world's top four satellite navigation systems. By 2012, the BeiDou system had covered the entire Asia-Pacific region. By 2020, it should be capable of covering the whole earth.

## Why was BeiDou developed?

Why was BeiDou developed when the GPS system already existed? One Austrian scholar gave an explanation at a navigation forum: "A country's navigation application could only achieve independence by having its own navigation system, which would be more reliable."

GPS is a satellite navigation system focusing on military use, with its ownership, rights of control and operation belonging to the US Department of Defence. A GPS sends out different wireless signals for military and civil use. Before 2000, the US military caused interference to the civil signal released by GPS, with the purpose of lowering the civil signal's precision and protecting the US's information security. Because control is in the hands of the US military, it causes its users in many countries to worry that the US military could suspend the civil signal at any time.

The BeiDou navigation satellite system could promise a continuous civil signal. For many countries, it is a double insurance. No matter whether it is for military or civil use, no matter whether it is for developing the economy or upholding state security, the navigation system will play an increasingly large role in future. In areas concerning a country's security, such as banking, transportation, public security, a home-developed system allows more guarantees.

## Knowledge link

### Chronicle of events for BeiDou navigation satellites

- On 3 February 2007, China successfully launched a BeiDou navigation test satellite at Xichang Satellite Launch Centre, entering its orbit precisely.
- At 04:11am on 14 April 2007, China successfully sent a BeiDou navigation satellite into space by the Long March-3A carrier rocket at Xichang Satellite Launch Centre. This BeiDou navigation satellite (COMPASS-M1) was the first in China's BeiDou navigation satellite system (also called COMPASS) construction plan. It flies in the medium earth orbit at an altitude of 21,500km. Its successful launch marked a new development phase for China's self-developed BeiDou navigation satellite system.
- At 00:16am on 15 April 2009, China sent the second BeiDou navigation satellite into a predetermined orbit by Long March-3C carrier rocket at Xichang Satellite Launch Centre. This Beidou navigation satellite (COMPASS-G2) was the second network satellite in China's BeiDou navigation satellite system construction plan. It was a geostationary satellite.
- At 00:12 on 17 January 2010, China successfully launched the third BeiDou navigation satellite by Long March-3C carrier rocket at Xichang Satellite Launch Centre.
- At 23:53 on 2 June 2010, China successfully sent the fourth BeiDou navigation satellite into predetermined orbit by Long March-3C carrier rocket at Xichang Satellite Launch Centre.
- At 05:30 on 1 August 2010, China successfully launched the fifth BeiDou navigation satellite, and sent it into predetermined orbit by Long March-3A carrier rocket at Xichang Satellite Launch Centre. The third BeiDou network satellite launched by China since 2010 was an IGSO (Inclined Geo Stationary Earth Orbit) satellite.
- At 00:26 on 1 November 2010, China successfully sent the sixth BeiDou navigation satellite into space by Long March-3C carrier rocket at Xichang Satellite Launch Centre. In this launch, China Satellite Navigation System Management Office put the BeiDou logo onto the carrier rocket for the first time.
- At 04:20 on 18 December 2010, China successfully sent the seventh BeiDou navigation satellite into predetermined transfer orbit by Long March-3A carrier rocket at Xichang Satellite Launch Centre.
- At 04:47 on 10 April 2011, China successfully sent the eighth BeiDou navigation satellite into predetermined transfer orbit by Long March-3A carrier rocket at Xichang Satellite Launch Centre. It was an IGSO satellite. This was the first launch of a BeiDou network satellite in 2011. This successful launch marked the completion of the basic system of the BeiDou's regional navigation satellite system, as well as China's satellite navigation system construction entering a new development phase.
- At 05:44 on 27 July 2011, China successfully sent the ninth BeiDou navigation satellite into predetermined transfer orbit by Long March-3A carrier rocket at Xichang Satellite

Launch Centre. It was the fourth IGSO satellite for the BeiDou network. This successful launch meant another solid step forward in the construction of China's BeiDou navigation system.

- At 05:07 on 2 December 2011, China successfully sent the 10th BeiDou navigation satellite into predetermined transfer orbit by Long March-3A carrier rocket at Xichang Satellite Launch Centre. This was the fifth IGSO satellite for the BeiDou network.
- At 00:12 on 25 February 2012, China successfully sent the 11th BeiDou navigation satellite into predetermined orbit by Long March-3C carrier rocket at Xichang Satellite Launch Centre.
- At 04:50 on 30 April 2012, China successfully launched the 12th and 13th BeiDou navigation satellites, which entered into predetermined orbit by Long March-3B carrier rocket at Xichang Satellite Launch Centre. This was China's first launch of two navigation satellites with one rocket. It was also the first time that China had launched two medium-and-high-earth-orbit satellites in this way.

## Large aircraft

### *Interpretation of key terms*

The large aircraft project was a major strategic decision of the Central Party Committee and the State Council to enhance China's innovation ability and strengthen its core competitiveness. It was one of 16 major special projects listed in the Outline of the National Programme for Long- and Medium-Term Scientific and Technological Development (2006-20).

The objective of the airliner project was to successfully develop a large passenger aircraft and the corresponding commercial power generator, airborne system and equipment and materials; to gain the ability to independently develop the market, design, manufacturing, trial flight, airworthiness and customer services; to realise the large-scale production of large passenger airliners and achieve commercial success.

On 13 March 2008, the State Council officially approved the establishment of the Commercial Aircraft Corporation of China, to plan the development of the trunk airliner and regional airliner as a whole.

### *Checking the past and present of the C919*

C is the first letter in both China and COMAC, the acronym of the Commercial Aircraft Corporation of China. It also has an implied meaning, which is to ascend into the international large passenger aeroplane market and to rival Airbus and Boeing in the international large passenger jet manufacturing industry, just like the letters of A, B and C.

The first "9" implies "forever" in Chinese culture, while "19" means the maximum passenger capacity of China's first large passenger aeroplane is 190 seats. The model following the C919 would probably be named the C929, in which "29" means the maximum passenger capacity of this model is 290 seats.

The C919 large passenger aircraft is a type of short-to-medium-range commercial trunk aircraft, for which China has the proprietary intellectual property rights. The C919 adopts the physical layout in which there are low-wing, twin-fan jets, a conventional horizontal stabiliser and retractable tricycle landing gear. It emphasises four features (safety, economy, comfort and environmental friendliness) and three "reductions" (weight reduction, drag reduction and emission reduction). Composite material accounts for 15% of whole aircraft and aluminium lithium alloy accounts for a further 15.5 % of the total. It also uses the most advanced engine and airborne system in the world. While based on the domestic market, the C919 also explores the international market. Its all-economy class layout comprises 168 seats, and the hybrid class layout 156 seats. Its first flight is expected to take place at the end of 2015, with airworthiness and first deliveries scheduled for late 2016.

## Knowledge link

### Development history of the C919

- In 2008, COMAC officially launched the project demonstration of the large passenger jet programme. With the strength and wisdom of the whole nation, it invited 468 experts from 47 domestic and foreign units to form a joint engineering brigade for the large passenger jet project, established an expert consultation team comprising 20 academics and experts, formed an initial general technological plan, completed a techno-economic feasibility research report for the project and arranged 14 special technology key projects that needed to be launched as the first batch.

Song Zhenping, Xinhua News Agency

**The C919 model made by COMAC attracts many visitors at the 2011 Asian Aerospace International Expo (photo taken on 8 March)**

- In May 2009, AVIC Chengdu Aircraft Industrial (Group)/AVIC Chengfei Commercial Aircraft Company (AVIC-CAIC/AVIC-CCAC) dispatched a joint design panel to the Shanghai Aircraft Design and Research Institute to participate in the design of the engineering prototype structure of the C919's nose section.
- On 19 June 2009, the AVIC-CAIC/AVIC-CCAC completed the assessment and unveiling of the general scheme for the manufacturing process of the engineering prototype.
- On 31 August 2009, the entire 3D mathematical model of the C919 engineering prototype was distributed.
- At 09:30 on 1 September 2009, the nose section of the C919 model began to be constructed at AVIC Chengdu Aircraft Industrial (Group). It was 7.9 metres long, 3.96 metres wide and 4.16 metres high and had a radar dome, cockpit, toilet, kitchen and cabin.
- On 8 September 2009, a model of the C919 was unveiled at the Asian Aerospace International Expo in Hong Kong. This was also the debut of the domestically made C919.
- On 24 November 2009, the assembling state of the upper and lower part of the C919 nose section passed assessment, and the joining of the two parts was realised.
- On 8 December 2009, the C919 engineering prototype went off the production line. The whole engineering design and manufacturing only took six months.
- On the morning of 10 December 2009, the acceptance ceremony for the backbone structure of the nose section of the C919 large passenger aircraft was held at AVIC Chengdu Aircraft Industrial (Group).
- On 21 December 2009, AVIC Commercial Aircraft Engine (ACAE) and CFM International (France) signed a memorandum of understanding in Beijing, planning to set up a final assembly and test run production line for the LEAP-X1C aircraft engine, which was used for assembling the C919 through a joint venture.
- On 25 December 2009, the backbone structure of the nose section of the C919 large passenger aeroplane was officially delivered in Shanghai.
- On 16 August 2010, a ceremony was held at AVIC-Hongdu (AVIC Hongdu Aviation Industry Group) in Nanchang, Jiangxi province, to mark the riveting of the first ISO-straight sections for the COMAC C919 jet-liner. This meant that the fuselage section of the C919 would enter the manufacturing phase. On the morning of that day, the Nanchang Aviation Industry City programme, which is for the manufacture of the C919's fuselage section, was started. At the same time, Jiangxi Hongdu Commercial Aircraft Corporation was also established to shoulder the task of developing the front, middle and rear fuselage sections of the C919.
- On 16 October 2010, the first sample piece of the C919 pressure bulkhead section went off the assembly line in Shenyang. It was the first large section sample of composite materials finished in the development of the C919. With a complex structure, it adopted brand-new materials and structure form. The success of its development further verified relevant engineering designs and manufacturing technique schemes, accumulating precious experience for the application of composite materials and technologies in the C919 project. It was a key technological achievement for the C919 large passenger aircraft project.
- On the afternoon of 15 November 2010, COMAC held the opening ceremony for the C919 large passenger aircraft demo mock-up in Zhuhai. With widespread attention at home and abroad, the 1:1 C919 large passenger aircraft demo mock-up was unveiled at Zhuhai air show.

- On the morning of 16 November 2010, COMAC signed the C919 aircraft launch customer agreement with six domestic and foreign enterprises – Air China, China Eastern Airlines, China Southern Airlines, Hainan Airlines, CDB Leasing and the US-based GECAS (GE Commercial Aviation Services), in which China has proprietary intellectual property rights. COMAC obtained 100 orders. This was confirmation of the first batch of C919 orders.
- On 17 December 2010, a sample piece of the horizontal stabiliser section for the C919 was successfully rolled out.
- On 24 December 2010, the roll-out ceremony of part of the aft fuselage section of the C919 was held at the No. 3 Academy of China Aerospace Science & Industry Corporation (CASIC).
- On 26 December 2010, the roll-out ceremony for parts of the aft fuselage section pylon sections of the C919 was held at AVIC SACC. The part of the aft fuselage section used 100% composite materials, with an advanced and domestic-pioneered designing scheme. The sample piece of the pylon section was the main load-bearing structure of the C919, taking the wing-mounted form.
- On 28 December 2010, the roll-out ceremony of samples of the central wing section, flap and moving mechanism section of the C919 aircraft was held at AVIC XAIC. The successful development of the two sections was of great significance in resolving the difficulties of the new technologies, new materials and new techniques, as well as in the management of airworthiness. The success also laid a solid foundation for follow-up experiments and verifications.
- On 19 April 2011, a strength test for the pylon section of the C919 was carried out at Nanjing Aeronautics and Astronautics University.
- On 20 June 2011, the 49th Paris Air Show was inaugurated at Bourget International Exhibition Centre. The 1:1 sample of the C919 was shown in front of an overseas audience for the first time.
- On the afternoon of 19 October 2011, ICBC Leasing and COMAC signed an agreement to order 45 C919 large passenger aircraft, bringing to 145 the total number of launch orders for the C919.
- On 21 October 2011, Sichuan Airlines signed an agreement with COMAC to order 20 C919 large passenger planes at signing ceremonies of both the memorandum of cooperation and the "trip to Sichuan by centrally-administered SOEs" cooperation programme between SASAC (State-owned Assets Supervision and Administration Commission of the State Council) and Sichuan provincial government. This marked the fact that the C919

Chen Xiaowei, Xinhua News Agency

**On 8 December 2011, the leaders of both COMAC, the company developing the C919, and CALC (China Aircraft Leasing Company) pose for a photo after signing an order agreement for 20 C919 large passenger aircraft in Hong Kong**

large passenger aircraft has received western China aviation market acceptance, with the total number of launch orders reaching 165.

- On 23 November 2011, Bank of Communications Financial Leasing signed an order agreement for 30 C919 large passenger aircraft with COMAC in Shanghai, bringing the total number of launch orders to 195.
- On 8 December 2011, China Aircraft Leasing Company (CALC) signed an order agreement for 20 C919 large passenger aircraft with COMAC, bringing to 215 the total number of launch orders.
- On 9 December 2011, the C919 passed state-level initial design evaluation and entered the detailed design stage.
- On 19 December 2011, the commencement ceremony of the first piece of sections for the COMAC C919 project was held at the CNC Machining Centre of the AVIC-CAIC. At 09:19, the milling of No.6 beam of the C919 large passenger aircrafts' front boarding gate was officially launched, marking the start of C919 component manufacturing and the forthcoming roll-out of the manufacture of the fuselage section for the C919 project.
- On 14 February 2012, the opening day of the Singapore Air Show, BOC Aviation signed an order agreement for 20 C919 large passenger aircraft with COMAC, making the total number of launch orders reach 235.

(A summary of information from www.comac.cc,
the website of the Commercial Aircraft Corporation of China)

# Chapter 2

# Electronic information

Over the past decade, China's electronic information industry achieved a series of innovative results. Since the start of the 12th five-year plan period, new industries have emerged such as the internet, cloud computing and the mobile internet, B3G (beyond third generation).

In the internet of things, the NDRC (National Development and Reform Commission), in collaboration with other departments, launched 12 national application exemplary projects in key areas, such as an intelligent transportation system, intelligent public safety management, intelligent agriculture, intelligent environmental protection and intelligent forestry, so as to coordinate the R&D of core technologies, set standards and industrialisation.

In cloud computing, the NDRC, in collaboration with the Ministry of Industry and Information Technology, gave a pilot demonstration of innovative developments, to promote cloud computing services, explore the feasibility of cloud computing, serve economic and social development, and guide and promote the development of cloud computing in an orderly manner. Based on this, the NDRC organised and implemented a group of cloud computing exemplary projects in five pilot cities, together with the Ministry of Finance and the Ministry of Industry and Information Technology.

In the next-generation internet (NGI), exemplary projects were actively carried out during the 12th five-year plan, which gave the road map, timetable, main task and policy measures of China's NGI development.

Progress has been made in B3G and digital television. The Digital Terrestrial Multimedia Broadcast (DTMB) standard, for which China has independent intellectual property rights, has gained entry into countries such as Laos, Cuba and Cambodia.

In information network infrastructure construction, research has been started for the implementation scheme of a China broadband strategy.

For the development of core basic industries, policies to further encourage the development of software and integrated circuit industries were raised; the design project on an integrated circuit (IC) was carried out, and a group of IC design companies with international competitiveness was nurtured.

## New generation broadband mobile wireless communication network

### *Interpretation of key terms*

The New Generation Broadband Mobile Wireless Communication Network major project is one of China's key strategic tasks. The overall objectives of this project are to: break through key technologies and core chips, enlarge domestic and foreign markets, forge the industry chain and innovation chain, build a ubiquitous network with the new generation network as a backbone, broad wireless access (BWA) as support, wireless sensor network as a new means of acquiring information, and organic combination of all parts. The implementation of the special project will promote the international ascent of China's communication enterprises and realise a great leap forward from a "large market" to a "great power" in the communication industry.

During the 11th five-year plan, on the basis of TD-SCDMA, China acquired a group of core technologies, which make LTE-Advanced TDD model, the new generation mobile communication technology for which China has independent intellectual property rights, becoming a mainstream standard internationally and strengthening the completion of backbone enterprises in international communication markets, such as China Mobile, ZTE and Huawei.

TD-SCDMA has achieved remarkable industrialisation results, forming a complete industry chain. With domestic enterprises as the core, the industry chain covers chips, terminals, systems, instruments and applications, involving more than 200 domestic and foreign enterprises. TD-SCDMA products are maturing with increasingly rich terminal products.

TD-SCDMA has started to reverse the situation of "big but not strong" for China's mobile communication industry. It has also saved patent fees for the industry, lowered prices of mobile communication products, nurtured talent in technology research, promotion of international standards, product R&D and project management, formed talent teams oriented towards industrialised development and promoted the development of China's communication industry.

The R&D of TD-LTE has steadily progressed. Its initial R&D achievements were displayed at Shanghai World Expo.

The 4G proposal by China has been identified as an international standard, with the proportion of independent intellectual property rights increased.

## *Sci-tech serving the people*

### China Mobile's TD covers 80% of urban areas

Zhu Zheng, Xinhua News Agency

**On 21 May 2009, staff display a mobile video phone using TD-SCDMA technology**

On 23 May 2012, Xi Guohua, chairman of China Mobile, said that, in 2011, China Mobile had fundamentally realised continuous coverage of the TD network in domestic prefecture-level cities, county-level cities and the main urban area of county towns, covering 80% of the urban area.

At present, China is at a crucial stage in the mass commercial application process, where user scale experiences explosive growth and the industry is moving in a virtuous cycle. The industry shows rapid growth momentum. Meanwhile, China is deepening testing of 4G TD-LTE technology with independent intellectual property rights owned by China.

With 80% coverage of urban areas in 2011, the number of TD users is steadily growing, with a smooth completion of the first-period construction task of TD-LTE scale test network in six cities: Shanghai, Nanjing, Hangzhou, Xiamen, Guangzhou and Shenzhen.

### 4G approaching, high-speed access no longer a dream

While wireless internet access has been chosen by more and more people, poor connecting speeds have been a problem for users. 4G brings more possibilities to people's lives.

Li Zhong; Released by Xinhua News Agency

**On 19 April 2012, at the 6th China Hangzhou Electronic Information Expo, an expo worker walks past the 4G wireless network application exhibition area**

World Telecommunication Day falls on 17 May. At KingKey 100 Plaza in Shenzhen, many citizens stopped at an exhibition booth to experience 4G, and found out what changes 4G would bring to their lives. 4G applications promise important benefits and new features such as high-speed internet access, high-definition video call, high-definition video monitoring, and latest 4G terminals such as the 4G express card, and mobile hotspot.

4G has a connecting speed 20 times quicker than 3G, and can transmit high-definition video images in real-time synchronisation. It can also support high-definition video display. One computer could open four windows at the same time broadcasting 4 HD movies.

Through the 4G network, participants at an HD video conference could even look into each others' eyes, and get a clear view of each others' expressions. Even if you're a million miles away from each other, it's like you're sitting in the same conference room.

Currently, the technology is still in the experimental phase. But China Mobile is accelerating the construction of a larger 4G test network in Shenzhen. For this purpose, 3,000 4G communication stations will be built, with an investment of Rmb1.2bn.

Core districts of Shenzhen such as Futian and Luohu already have 4G network coverage now.

Under the support of 4G, wireless city businesses integrating various functions such as real-time traffic images under the current 2G network would become smoother and more vivid.

Currently, wherever there is a cellphone signal, citizens could use mobile phones to handle enquiries and payments such as social security, public accumulation funds, utilities and gas bills. They can also use mobile phones to check information about port customs clearance, traffic videos and employment. Hospital appointment registrations and cinema ticket bookings can also be done by mobile phone. All these have brought more convenience to citizens' lives.

As mobile internet technology is applied more in city management and in the lives of residents, "wireless city" is becoming a bond connecting social management, government services and all walks of life.

## Next generation internet

### *Interpretation of key terms*

CNGI, the demonstration project for China's next generation internet, is a starting project to implement the development strategy for China's next generation internet. In December 2003, the NDRC (National Development and Reform Commission) approved a feasibility study report of the core network construction programme for CNGI. The programme's main purpose was to establish a test platform for the next generation internet, with IPV6 as the core.

According to the development goals of the next generation internet in the 12$^{th}$ five-year plan, as well as the road map and timetable released by departments such as the NDRC and the MIIT (Ministry of Industry and Information Technology) on 29 March 2012, internet penetration would reach 45%, accelerating the integration of the three networks, and the number of IPV6 broadband users would exceed 25m. The interflow of the main businesses of IPV4 and IPV6 would be realised. The amount of IPV6 address requests could fully satisfy the needs of users.

### *Focusing on IPV6*

IPV6 is the abbreviation of Internet Protocol Version 6. It is the next generation IP protocol designed by IETF to take the place of the current IP protocol (IPV4). Currently, the global internet uses the TCP/IP protocol family. IP is the network layer protocol of the protocol family, its core protocol. IPV6 is being continuously developed and improved. In the near future, it will take the place of IPV4, which is currently in wide use. Everybody will own more IP addresses.

#### China's internet leaps into the IPV6 age

At the end of 2011, the State Council published a road map and main goals of the next-generation internet based on large-scale deployment and commercial use. On 29 March 2012, seven ministries and commissions including the NDRC and MIIT jointly issued some suggestions on how to develop NGI during the 12th five-year plan, giving an interpretation on this major strategic deployment. Through the guidance and drive of this suggestion, China's internet will hopefully be enhanced, entering the IPV6 age.

***How large is the next generation internet?***

According to data by experts, based on the current IPV4 internet, there are 4bn network addresses in the world. The next generation internet based on IPV6 has addresses whose number is two to the power 128, which is 1,029 times that of the current internet. One commonly-used pictorial saying is that the number could make "every grain of sand on earth has its own IP address".

By the end of 2011, the number of Chinese internet users had reached 513m, and the internet penetration was 38.3%. But China only owns around 332m IPV4 addresses (Hong Kong, Macau and Taiwan not included). Even if technologies such as NAT (Network Address Translation) postpone consumption of IPV4 addresses, rapidly increasing application demand is still not fulfilled. Furthermore, it can significantly increase network complexity and management difficulties, lower network and information security levels and service quality. During the 12th five-year plan, China's internet penetration rate would exceed 45%, with a more prominent address shortage problem. At present, the root name server has realised the support for IPV6. The distribution speed of IPV6 addresses by the global internet management institution is ever-accelerating. IPV6 has got the basis to be widely applied, pushing the internet to evolve from IPV4 to IPV6. Based on this, developing the next generation internet has become a global priority.

***How good is the next-generation internet?***

Our expectations of the internet are increasing. From email, browsing for news, instant communication to search, forums video, and the various demands of the mobile phone age, how good is the internet likely to be in future? Domestic telecommunication operators are badly in need of rich network address resources. Equipment manufacturing enterprises are badly in need of finding new growth points. Service-provision enterprises are badly in need of special services. Users are badly in need of advanced network facilities and safer business experiences with better quality. Given these needs, emerging interactive applications such as the internet of things, cloud computing, mobile internet and the integration of three networks will develop on a large scale.

In the "sixth age", the internet not only has IPV6, but also rich address resources, advanced facilities, energy-saving features and reliable safety. It can provide a larger amount of information and diverse business applications, more intelligent support connections, enabling it to lay "a more solid and stronger information foundation for social production and social living". For this purpose, the "suggestion" said that IPV6 broadband access businesses will be carried out, such as 3G mobile communication and follow-up evolution technology, optical fibre networks, Ethernet, WLAN (Wireless Local Area Network), so as to promote the commercial use of broadband data business based on IPV6; personalised interactive businesses that have considerable need for addresses, high speed and high mobility will be actively developed; a basic business platform for the integration of three networks based on IPV6 will be constructed, and the development of business applications about integration, such as mobile multimedia radio and television, IPTV, mobile TV and broadband internet access for digital TV, will be accelerated. The focus will be put on the internet of things, cloud computing and mobile internet, and the country will actively push forward applications for the next generation internet in key areas including education, agriculture, industry, medical services, transportation, railway, water conservation, environmental protection and social management.

***How will the next-generation internet be built?***

By the end of 2013, we had begun to carry out the small-scale commercial use pilot of the IPV6 network; the DNS (Domain Name Server) of telecommunication operators will support IPV6 visits and analysis; IPV6 will be supported by the businesses of telecommunication operators, and fixed and mobile terminal for internet access. From 2014 to 2015, the mobile internet business will transit to IPV6 comprehensively. New-type businesses such as the internet of things and cloud computing will adopt IPV6 addresses when IP network addresses are needed. The existing businesses of telecommunication operators will be moved to IPV6 gradually. Newly increased fixed and mobile terminals for internet access will support IPV6. According to the road map given by the "suggestion", telecommunication enterprises will shoulder an important task in the construction of the next generation internet. China will accelerate IPV6 upgrading and rebuilding for the public backbone network, metropolitan area network, IDC (Internet Data Centre), business system, supporting system; improve the performance of network equipment; accelerate IPV6 upgrading, rebuilding and large-scale deployment for public mobile, wired and wireless broadband access networks.

Meanwhile, the Chinese government will increase capital investment, drive the input of social capital and implement the major projects of the next generation internet during the 12th five-year plan in phases, actively guide links of the industry chain, such as telecommunication operators, cable TV operators, software R&D enterprises, equipment manufacturing enterprises and service providers to develop the next generation internet, so as to ensure the realisation of development goals.

Entering the sixth age, the public will have better application experiences. Information service workers will face stronger challenges and bigger opportunities. To have a head start in the new round of development, we should not only strengthen the inner power, but also make a concerted effort, and arouse vitality. (By Hong Liming)

(*People's Post and Telecommunications*, 5 April 2012)

## *Sci-tech serving the people*

### What will broadband China bring to the people?

On 9 May 2012, a State Council executive meeting passed the "Several Opinions on Vigorously Promoting Information Development and Effectively Protect Information Security", and determined that China will implement the "broadband China" project. So far, China's broadband project has ascended from enterprise and industry level to national level. What changes will the project bring to people's daily lives?

#### *Higher internet speed*

A State Council executive meeting pointed out that we will accelerate the upgrading of the information network, push forward the fibre-to-home effort, and realise the universal service of administrative villages.

Fibre-to-home means a qualitative leap in broadband speed. In the past, to promote this concept, Chinese telecommunication operators needed to make great efforts to negotiate with various stakeholders, including property management companies and developers. The "last 1km" problem of "fibre-to-home" was a long-standing headache for broadband construction.

"When the broadband project ascends to national strategic level, optical fibre could enter homes as an important public infrastructure, just like water, electricity and gas," said Shang Bing, Deputy Minister of MIIT.

According to Shang Bing, during the 12th five-year plan, China's optical fibre network will cover commercial buildings and newly-built communities. The fibre-to-home rate for newly built urban residential buildings will exceed 60%. Broadband access capacities in urban and rural areas will reach 20MB and 4MB respectively, with that of some developed cities reaching 100MB. The utilised bandwidth level for users will improve substantially. Meanwhile, the 3G network will cover urban and rural areas, realising continuous coverage of wireless broadband data businesses in hot areas.

Since 2012, China has accelerated broadband construction. As from 1 May, China Telecom officially introduced "Lighting up Optical Fibre Communities" in 21 southern provinces and some northern provinces. Access to each home will reach a maximum of 100MB. Users could become a China Telecom broadband subscriber by registering in its online business hall.

"How fast is optical broadband? It took four to five hours to download an HD movie using a 4MB broadband, but just over 10 minutes with China Telecom's 20MB optical network," said Wang Xiaochu, China Telecom's vice chairman.

***Lower fees***

Ms Jia lives in Weidong Xindu district of Jinan. When she came home from Germany in 2010, she had a 2MB ADSL broadband. Now it has been upgraded to 10MB for free, with only a Rmb40 monthly fee and 100 minutes of free calls. "After the network upgrade, connection is no longer lost on rainy days," she said.

"After the roll-out of fibre-to-home, China's broadband is bound to speed up but a user's monthly broadband fee will stay the same or even fall. This way, the flow price per MB will see a huge change," said Wu Hequan, an academic at the Chinese Academy of Engineering.

According to Wang Xiaochu, in 2012, China Telecom increased to 25m the number of fibre-to-home families, making the total exceed 55m. Wired broadband subscribers grew by 16m, increasing the number of subscribers beyond 100m, out of which 13m families are new internet subscribers through wired broadband access, making the total number reach 83m. This has helped national broadband penetration increase by 3.25%. For the whole year, more than 10,000 administrative villages will have broadband service.

Meanwhile, China Telecom plans to increase the proportion of subscribers using 4MB or faster broadband products go beyond 50%. For the whole year, WiFi hot spots will increase by 300,000, to 900,000.

**More integrated information**

A staff member from BII Group introduces a set of agricultural intelligent control systems based on IPV6

Zhao Wanwei, Xinhua News Agency

The State Council executive meeting said that the country would accelerate deployment of the next generation internet, develop key chips, equipment, software and systems for it. The government will accelerate the integration of the telecommunication, radio and TV networks, and internet, and nurture relevant industries and markets.

According to the introduction by Shang Bing, during the 12th five-year plan, China will formulate a strategic action plan for the next generation internet (NGI) project. "China has begun to carry out new-type NGI business development, experiments of the current network, and online applications, support current businesses to move to IPV6, and focus on the promotion of industry applications in areas such as education, medical treatment, transportation, railways, water conservation, environmental protection, agriculture and industry."

Under the united deployment of the Broadband China strategy, China will accelerate the steps of three network integrations. During the 12th five-year plan, China will continue to push forward the double entering of radio and television and telecommunication businesses, issuing business licences to eligible enterprises. Relevant departments will organise pilot areas to assess the implementation effect, evaluate pilot business types, operation methods and supporting measures. Following the pilot studies and evaluation, the country will enlarge the pilot range and scope gradually. It will also improve engineering standards and regulations, promote the joint construction and sharing of telecommunication networks and radio and television network infrastructure, enhance networks' comprehensive benefits, and guide the development of integrated businesses such as IPTV and mobile TV.

"The final outcome of the broadband project is that it not only benefits the people's informational life, but also promotes the process of society informatisation, and enhances the operation of all economic and social domains," said Wang Xiaochu.

## CHI major special project

### *Interpretation of key terms*

The major project of core electronic devices, high-end general-purpose chips and infrastructural software products is called CHI Major Special Project. It is one of the 16 major sci-tech special projects identified in the Outline of the National Programme for Long-and-Medium-Term Scientific and Technological Development (2006-20). After the implementation of the CHI Major Special Project in the 11th five-year plan, in the areas of core electronic devices, high-end general-purpose chips and infrastructural software products, through continuous inputs and technological innovation, China has overcome a series of key technologies and developed a group of core strategic products. The R&D of core electronic devices achieved a major breakthrough, and has begun to be applied in several areas; R&D on high-end generic chips has narrowed the gap with an internationally advanced level; basic software products have formed a relatively mature industry chain in some areas.

The petaflop computer Tianhe-1, which is independently developed by China and partially adopts domestic high-performance multi-core CPU FT-1000 and Kylin operation system, ranked first place among the top 500 world super computers in 2010; the embedded application platform of the smart phone now competes against foreign products. The high-performance digital signal processor and HDTV PDP Display Driving Chip have been successfully developed, breaking the monopoly of foreign manufacturers; high-performance embedded CPU chips have realised large-scale applications; a real-time embedded operation system has realised applications in several auto factories' own-brand cars; the advanced EDA tool platform has realised mass popularisation, filling several gaps in domestic EDA tools; Chinese office software and domestic middleware have achieved important progress; domestic basic software has been applied in the pilot project of the smart dispatching system of the power grid.

### *Result highlights*

- On 17 April 2011, China's first PCRAM chip gaining independent intellectual property rights, was successfully developed, breaking foreign countries' monopoly on memory chips' production technology. This PCRAM test chip has an 8MB memory capacity.

Each memory chip on the 8-inch silicon slice has a more than 99% rate of finished products for memory cells.

- The FT-1000 CPU is a multi-core 64-bit high-performance CPU that has successfully developed under special support. Its performance in actual measurement has reached the international commercial use CPU level. 2048 slices of FT-1000 were integrated in Tianhe-1, which ranked first place among the top 500 world super computers in November 2010. It has been applied in several areas such as weather forecasting, oil exploration, financial analysis and remote sensing data processing.
- The Kylin server operating system is a 64-bit server operating system independently developed under special support. The system broke through key technologies such as hardware adaptation, kernel protection and resource deployment. The system has passed the military's B+ security certificate, ISO Linux Standard Base 4.0 certification and the GB18030 Chinese compliance certification. It has been successfully deployed in Tianhe-1, and got successful applications in several key equipment systems.
- The ShenWei processor SW1600 is a multi-core 64-bit high-performance CPU that has been successfully developed under special support. It adopts several key technologies such as an independent instruction set, SoC integration, integrated memory controller, and standard I/O interfaces, with a peak calculation speed at 140.8bn times per second.
- The Loongson CPU is a multi-core 64-bit high-performance generic CPU that has been successfully developed under special support. It has been applied in products such as the blade servers and high-performance computers.
- The successful development of the HR-1 high-performance DSP chip bears a significant meaning for breaking the foreign blockade on techniques, enhancing China's sustainable development capability in developing equipment, solidifying national security, and improving China's innovation ability in this domain.

# High-performance computer and grid service environment

## *Result highlights*

### Five high-performance computing systems

- The calculating speed of Tianhe-1 reached 2.57 PFlops, winning first place in the world.
- The calculating speed of the "Dawning Nebula" reached 1.27 PFlops, ranking third place in the world.
- The calculating speed of the Sunway BlueLight MPP reached 1.10 PFlops, built with a domestic CPU.
- The calculating speed of the Dawning 5000A reached 180.60TFlops, ranking 10th place in the world in 2009.
- The calculating speed of the Lenovo Deepcom 7000 reached 102.80TFlops, ranking the 19th place in the world in 2008.

- The CNGrid service environment was built up, including 11 nodes on the mainland and in Hong Kong, with an aggregate calculating capacity of 458.1TFlops, and a storage capacity of 2.94PB
- Developing CNGridGOS4.0, China's national network software with its own intellectual property rights
- Deploy more than 230 application software and tool software applications
- Opening up two grid application communities: scientific computing and industrial simulation and optimal design
- Supporting the research of more than 700 "973", "863", National Science Foundation programmes, and other important engineering projects
- Supporting applications in 16 areas, including drug discovery, aircraft design, railway freight design, weather forecasting and oil exploration, and achieving key results
- Accelerating the development process of new drugs, with virtual screening results of drugs targeting the influenza virus
- The GRAPES meteorological model successfully forecasted Typhoon Sinlaku
- Realising the sharing of freight data resources at 19 national railway nodes
- Realising the rapid processing and parallel computing of ecological monitoring information for the Three North Shelterbelt programme
- Recurring the evolution process of matter distribution in 5bn cubic light years
- Realizing ten-thousand-core parallel simulation of key and complex flow problems in aircraft design

## National supercomputing centre in Jinan unveiled, domestic CPU and software adopted

On 27 October 2011, The National Supercomputing Center in Jinan was officially unveiled. This was China's first petaflop super-computing system consisting of 100% domestic-made CPU and system software. It showed that China was capable, after the US and Japan, of building a petaflop super-computer with its own CPU.

The Sunway BlueLight MPP computer system installed at the National Supercomputing Jinan Centre was developed by the National Parallel Computer Engineering Technology Research Centre. The system has 8704 ShenWei SW1600 processors developed by the National High Performance IC (Shanghai) Design Centre, with a peak performance of 1.07016 PFLOPS, a sustained performance of 796 TFLOPS, a performance per watt of 741 megaflops and a LINPACK efficiency of 74.4%. Its packing density, performance per watt and general proficiency reached world-class levels. The Jinan Centre adopts 100% domestic CPU and system software, realising the independent control of core technologies of the national key information infrastructure.

Xu Suhui, Xinhua News Agency

On 27 October 2011, the National Supercomputing Centre in Jinan was officially unveiled. Staff check the host section of the Sunway BlueLight MPP computer system

China has built up three petaflop super-computing centres in Tianjin, Shenzhen and Jinan.

## *Focusing on Tianhe-1*

### Tianhe-1 leads world ranking

On 17 November 2010, the 36th edition of the World's Top 500 Supercomputers list was officially published. The Tianhe-1A system at the National Supercomputer Centre in Tianjin was ranked the world's fastest machine, with 2.566 PFlop/s Linpack Performance and Rpeak of 4700 TFlops/s.

### Interpreting Tianhe-1's application results

Figure 1: Tianhe-1 has an Rpeak of 4700 TFlops/s, with a Linpack performance of 2.566 PFlop/s. In November 2010, the system ranked first in the 36th edition of the World's Top 500 Supercomputers list.

Figure 2: The number of subscribers has reached 300, with its application covering many key areas such as oil exploration, biological medicine, aerospace,

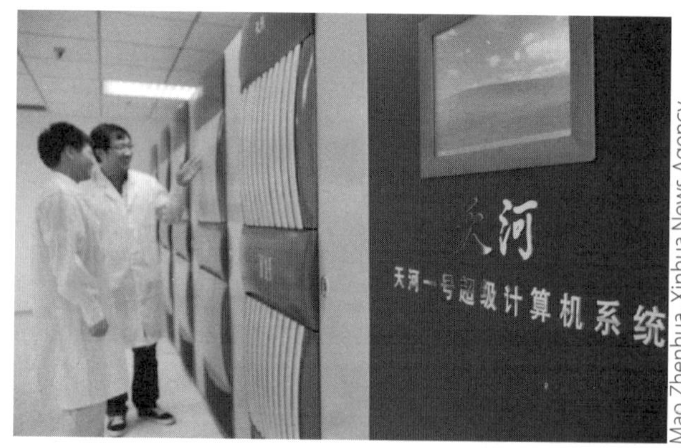


Mao Zhenhua, Xinhua News Agency

Researchers at the National Supercomputing Centre in Tianjin check the Tianhe-1, which has completed assembly

high-end equipment manufacturing, civil engineering design, weather forecasting, marine environment, new energy, new materials, basic scientific research, animation, and film and television rendering.

Figure 3: Build up five application platforms, including oil exploration data processing, biological medicine data processing, animation and film and television rendering, high-end equipment manufacturing, product design and simulation and geographic information.

Figure 4: In oil exploration, up to 7,100 nodes are used to complete the complex 3D processing task of 70,000 explosions of oil seismic exploration data covering 1,060 sq km within 16 hours.

Figure 5: Develop cloud computing platform used for animation, film and television rendering. The production cycle of animation, film and television works is shortened from four-to-six months to several days, capable of providing rendering for eight works at the same time.

Figure 6: Finish the high throughput virtual screening for the development of new medicines at high speed. Traditionally, research of a new medicine takes more than 10 years, costing around US$1bn; the high throughput virtual screening will shorten the R&D cycle by about 18 months, and reduce costs by more than US$100m.

## Knowledge link

### Chronicle of events on China's supercomputer development

- In 1983, China's first giant-scale computer, named the Galaxy and capable of 100m calculations per second, was born at the National University of Defence Technology. China became the third country capable of independently designing and manufacturing a giant-scale computer, after the US and Japan.
- In 1992, the National University of Defence Technology developed a general parallel giant-scale computer Galaxy-II, with a peak speed of 1bn calculations per second. It was mainly used for medium-term weather forecasting.
- In 1993, the National Research Centre for the Intelligent Computing System, which established Beijing Sugon Computer Co, successfully developed the Shuguang-1 holohedry shared memory multi-processor. It was China's first parallel computer developed by means of general microprocessor chips on the basis of a super-large scale integrated circuit and standard UNIX operating system.
- In 1995, Sugon launched the Shuguang-1000, with a peak speed of 2.5 gigaflops and a real calculation speed of 1 gigaflops. Shuguang-1000 has a similar structure and technology to large-scale parallel computer system launched by Intel in 1990. It reduced the gap with similar foreign products to about five years.

- In 1997, the National University of Defence Technology successfully developed Galaxy-III, a 10 gigaflop giant-scale parallel computer system, with a peak performance of 13 gigaflops.
- From 1997 to 1999, Sugong launched the Shuguang-1000A, Shuguang-2000I and Shuguang-2000II super processors, with a peak calculation speed of more than 100 gigaflops.
- In 1999, the National Research Centre of Parallel Computer Engineering & Technology (NRCPCET) developed the Sunway-I computer, with a peak calculation speed of 384 gigaflops. It was put into use in the National Meteorological Centre.
- In 2004, the Shuguang-4000A, jointly developed by the Institute of Computing Technology, Chinese Academy of Sciences, Sugon and Shanghai Supercomputing Centre, reached a calculation speed of 10 TFlops.
- In 2008, DeepComp 7000 was China's first heterogeneous cluster computing system to break the speed of 100 TFlops, with the Linpack exceeding 106.5 TFlops.
- In 2008, Shuguang-5000A reached a peak speed of 230 TFlops, and the Linkpack 180 TFlops. The Shuguang-5000A could perform various large-scale scientific engineering computing and business calculations.
- On 29 October 2009, China's first PFlops super computer Tianhe-1 was born. It had a peak speed of 1206 TFlops/s, with a Linpack performance in actual measurement of 563.1 TFlop/s, making China the second country capable of developing a PFlops super computer after the US.
- On 16 November 2010, the 36$^{th}$ edition of the World's Top 500 Supercomputers list was officially issued. The Tianhe-1A system at the National Supercomputer Centre in Tianjin was ranked the world's fastest computer, with 2.566 PFlop/s Linpack Performance and an Rpeak of 4700 TFlops/s.

## Integration of three networks

### *Interpretation of key terms*

The integration of three networks means that, in the evolution of telecommunication network to broadband communication network, radio and television network to digital TV network, and internet to the next generation internet, the three networks have increasingly similar technological functions and business domains through technological upgrading. The interconnected network and shared resources could provide users with many services such as voice, data, and radio and television. The integration doesn't mean the physical unification of the three networks, but instead the integration of high-level business applications. The integrated networks have wide applications, including smart transportation, environmental protection, government affairs, public security, and safe housing and accommodation. The mobile phones of the future could be used to watch television and surf the web. Television could be used to make phone calls and surf the web. And computers could be used to make phone calls and watch television. The interlocked networks form a structure of, "There is a bit of me in you and a bit of you in me".

## Knowledge link

### China's integration of three networks

- On 7 October 2008, MIIT (the Ministry of Industry and Information Technology) issued the Emergency Notice of Promoting Joint Construction and Sharing of Telecommunication Infrastructure, requiring preparatory groups of China Telecom, China Mobile, China Unicom and the OFCOMS of provinces, autonomous regions and municipalities to jointly construct and share existing equipment so as to reduce duplication.
- At the end of October 2008, Dai Xiaohui, deputy director of the Department of Science and Technology at the MIIT, said that relevant national departments are actively creating good conditions, giving a clear direction for a video-oriented development path in the 3G age through formulating policies, measures and common standards to regulate new business development after the integration of three networks.
- In November 2009, Liu Lihua, Party Member of the MIIT, pointed out at the 8th Information Port Forum that the MIIT will further strengthen coordination and communication with relevant departments, actively explore the mechanism and path to realise the integration of the three networks, accelerate the process in the aspects of business, network and terminal, to nurture new growth points. But telecommunication operators should grasp the trends of communication technology towards mobility, broadband, video and IP, exploit the advantage of all-business operation, strengthen businesses, technologies and network innovation and vigorously promote the integrated development of a fixed network, mobile network and the internet.
- On 13 January 2010, a Sate Council executive meeting decided to accelerate the integration of the telecommunication network, radio and television networks and the internet, realise the integration of the three networks within five years and clearly raise phased objectives and key works.
- On 30 June 2010, after assessment and approval by the State Council's coordination panel for the integration of three networks, the list of the first batch of pilot regions (cities) for the integration of three networks was determined. This marked the official launch of the piloting for integration. These areas and cities are Beijing, Dalian in Liaoning province, Shanghai, Nanjing in Jiangsu province, Hangzhou in Zhejiang province, Xiamen in Fujian province, Qingdao in Shandong province, Wuhan in Hubei province, Changsha-Zhuzhou-Xiangtan region in Hunan province, Shenzhen in Guangdong province and Mianyang in Sichuan province.
- In February 2011, Shenzhen IP TV's integrated broadcasting and control platform was built up smoothly and finished a link-up with CCTV's general platform, providing more than 100 SD live channels, 15 HD live channels, 20 Carousel channels, 20,000 hours of on-demand programmes and a broadcasting ability of six-hour time shifting, seven-day playback show. This laid a solid foundation to broadcast positive and healthy IP TV programmes.
- In February 2011, Sichuan IP TV integrated a broadcasting and control platform with CCTV's general platform, and received authorisation for commercial use. It was designed to meet the needs of pilot businesses for the integration of the three networks and laid a good foundation to broadcast positive and healthy IP TV programmes.
- On 8 May 2011, Wuhan Mobile and Wuhan Broadcasting & TV jointly signed a strategic cooperation agreement of the "Integration of Three Networks Co-build G3 Digital Family". The integration of three networks officially entered ordinary Chinese homes. The annual fee for the "family package of three networks' integration" was Rmb1,399,

through which customers could enjoy one year of 2M broadband, HD interactive TV and Rmb600 of telephone expenses.

- On 30 December 2011, the General Office of the State Council distributed a notice on second phase pilot areas/cities for the integration of three networks. There were 42 places on the list, including Tianjin, Chongqing, Shijiazhuang and Xi'an.

## *Sci-tech serving the people*

### Integration of three networks benefiting people's lives

#### *One line does three things*

Outlook of Three Networks: After dinner, if you want to watch a World Cup football game played last night, you only need to switch on the TV, connect to the internet, choose the match you want, and enjoy. Then, a call comes in, with the caller number displayed on the TV screen; so, you pick up the remote control at your side, and take the call...

After the integration of the three networks, televisions and computers would be fused into one: you can watch TV on your computer and surf the internet on your TV. You can also make a phone call through your computer. In other words, after the integration of the three networks, consumers only need to choose one operator and install one network to watch TV, surf the internet and make phone calls at home, without extra wiring.

The "integration of the three networks" will revolutionise digital home use. For most users, though the integration hasn't yet been realised, they can already use an LCD TV as a computer screen, or change a computer screen into a TV through a TV tuner card. Once the integration occurs, TVs and computers will be connected seamlessly and integrated without the aid of other devices. Moreover, as the computer can make calls with landline and mobile phones, it will replace the current landline phone as the most powerful and affordable means of communication.

#### *Overall charges will fall*

Outlook of Three Networks: Each month, Ms Zhang has to pay many family bills. Her internet bill is Rmb88, telephone bill is Rmb70-80 and optical fibre bill is Rmb20. Added together, these bills are expensive. After the integration of the three networks, Ms Zhang only needs to choose a "three-network operator" and pay one account; then broadband, telephone and TV can be used at will....

Consumers will first choose which company they want after the integration of the three networks. Then the next consideration will be price.

Currently, bills for radio and television services, telecommunication services, and internet services need to be paid separately. But, after the integration of the three networks, users only need to pay once for using all three. Meanwhile, operators will provide different services, with users having more choice. Competition within the industry could be fierce, so charges are likely to go down, which could ensure real benefits for users.

Jin Liangkuai, Xinhua News Agency

**On 23 March 2011, the China International Radio & TV Information Network Expo was held in Beijing. A company representative displays a three-screen interactive system of computer, TV and mobile phone**

***More personalised and diverse services***

Outlook of Three Networks: after integration of the three networks, users will continue to enjoy the same services, but at a higher speed. In the near future, broadband will be calculated by 100MB units. If you are a moviegoer, you could choose to watch the online broadcasting of HD movies above 1080p (the p refers to progressive scanning) at any time. It takes only 30 seconds to download a game of 300MB.

After the integration of the three networks, networks will be interconnected, probably employing a unified IP protocol. Among the three networks, there will be mutual competition and cooperation. The three businesses of data, voice and image will be delivered through the same network and platform, with a trend towards providing a diverse, multi-media and personalised service.

For example, the double functions of TV and telephone in a computer could make ordinary people reduce costs. In future, it is probable that a watch-like mobile phone will be launched that can be worn on the wrist, and can be used to watch TV and surf the internet. This makes a terminal product provide diverse services and convenience. With the incoming of the “three networks age”, our lives will become better and better. (Huo Peng)

(*Tianfu Morning Paper*, 9 July 2010)

# Cloud computing

## Interpretation of key terms

Cloud computing is the fourth IT industry revolution after the mainframe computer, personal computer and the internet. It has changed the network application model, and will become the basis for driving the future growth of many industries, including IT, the internet of things and e-commerce, and promoting the overall upgrade of the information industry. The model advocated by cloud computing technology is aggregating all hardware (CPU, storage, network), software, data and services, and making individual and enterprise subscribers visit, share, manage and use resources through large-scale data and application centres, as well as the internet.

China stands at the same starting line on cloud computing as foreign countries. With nearly all software enterprises aware that cloud computing is an opportunity as well as a threat, cloud computing has begun to play an important role in information construction and application, pushing forward the upgrading of traditional and emerging industries.

### Knowledge link

#### China's cloud computing

- On 3 April 2010, China's first multi-system, multi-user cloud computing application network platform, www.worldhm.com, was officially unveiled in Beijing.
- From 21-22 May 2010, the second China Cloud Computing Conference was held in Beijing.
- On 9 July 2010, the Zhongguancun Cloud Computing Industry Alliance was officially established. Beijing announced that the city would strive to become a world-class cloud computing industry base with Rmb50bn in five years.
- On 9 November 2010, at the fourth China-US Internet Forum, He Hongbao, director of the Internet Research Centre of China Academy of Telecommunication Research of MIIT, said China's cloud computing was committed to providing services for 42m small and medium-sized enterprises, assisting them to step over the information threshold of capital, technology and talent.
- In April 2011, Shenzhen International Joint Laboratory of Cloud Computing was inaugurated. At the newly-established remote data centre, tens of thousands of computers and servers connected to each other to form a computer "cloud". Shenzhen International Joint Laboratory of Cloud Computing is a professional technological and application R&D laboratory created by domestic and foreign enterprises, such as Shenzhen Cloud Calculation Industry, Intel, IBM and Kingdee. In January 2012, the cloud-computing centre received approval.
- On 6 April 2011, Chongqing officially launched Liangjiang New Area's international computing centre and Chongqing Data Industrial Park. Liangjiang New Area can be

found at the Water and Soil High-Tech Industrial Park. It is planned to invest Rmb40bn, covering an area of 2.07m sq metres, including a data centre of 3 sq km that complies with international standards. Chongqing Data Industry Park was created by China International E-commerce Centre, which is investing Rmb1.6bn. It is the first project to enter Liangjiang international cloud computing centre.

- From 18-20 May 2011, the third China Cloud Computing Conference was held in Beijing, with the theme of Application Road of Cloud Computing.
- On 24 May 2011, Shandong Cloud Computing Industry Alliance was established at Shandong province's computing centre. Forty enterprises and public institutions engaged in cloud computing industry research and production became members of it.
- On 21 July 2011, 35 high-tech enterprises formed a cloud computing industry alliance, in Tianjin's Binhai New Area. Participants included Tianhe-1, HP, and Tencent.
- On 31 August 2011, China Telecom officially launched the Tianyi cloud computing solution. In 2012, China Telecom officially launched Tianyi cloud computing products such as cloud host and cloud memory. This was the first domestic telecommunication operator to release a cloud computing strategy and solution. It will further promote the development of the domestic and global cloud computing industry.

## *Sci-tech serving the people*

### Cloud computing is everywhere

An International Exhibition of Cloud Computing in Chongqing featured a dazzling variety of clouds, such as a medical and health cloud for patients to register and make enquiries at home and an e-government cloud to facilitate the search for livelihood information via mobile phones and other devices.

Cloud computing means the establishment of a powerful data centre for the purposes of user storage and calculation.

Just as households used to dig a well to source water, nowadays they simply need to turn on a tap. Take one ordinary Chongqing citizen, for example, who wants to experience the following "clouds":

**Medical cloud** Residents in 38 districts and counties of Chongqing want to get a health card within a year, and inquire about any past medical problems; medical insurance settlements will not be limited by area, and long-distance transactions can be made through a medical cloud for urban workers' medical insurance, urban and rural residents' medical insurance and commercial medical insurance. Residents could also use a computer, television or mobile phone to make a doctor's appointment, and have the medical cloud remind them to get healthcare services such as vaccinations, physical examinations and chronic disease check-ups.

**Government affairs wireless-cloud** Government affairs are under preparation. Information on industry and commerce, policing and emergency services could be shared through terminals. You only need a mobile phone to inquire about a variety of issues, and sort out affairs such as TV payments, electricity and gas inquiries. Currently, the Chongqing cloud-based intelligent city platform has a total number of 225,000 subscribers.

**Transportation cloud** When citizens go to work or on holiday, they can use mobile phones or a navigator to check the transport situation, choose the best route and find a free parking space. "Clouds" can also be found when travelling. Currently, the transportation cloud can be used in five scenic spots including Fengdu Ghost Town, Wushan Red Leaves, Qianjiang Xiaonanhai, Tongnan Rape Flower and the New North Zone Happiness Square.

Citizens can also visit more than 100 cloud communities in the city, and inquire about community announcements, large supermarkets and public bus routes. In addition, the transport safety problem can also be addressed through the cloud. Wu Jianzhong, director of the computer committee at the China Railway Society, said that there had been severe traffic accidents before. In future, smart transportation will be built up gradually through the internet of things and the application of cloud computing. This will make transport safer and also allow for the prospect of blind people being able to drive.

**Colourful "cloud industry"**

The cloud industry is forming in the virtual world. Wang Zhentang, chairman of Acer, said commercial models, technologies and the integration of commercial services initiated by cloud computing provide opportunities for innovation, bringing a far-reaching impact on governments, enterprises and individuals. Liu Duo, vice director of the China Academy of Telecommunication Research at the MIIT, pointed out that cloud computing could drive industry development and reform. He identified two trends:

First, a clearer division of labour based on specialisation. Enterprises won't need to reconstruct their communication service system; instead, they could transfer this work to a communication company that specialises in information services.

Second, transfer of the whole value chain. In the age of cloud computing, the whole value chain will move to the cloud. All contents and calculations could be put into the cloud, but this will require stronger storage and processing capabilities.

After realising the integration of large-scale resources, the service cost of cloud computing will fall. It will be more convenient for many industries to use technology and information.

Technologies bring in industry changes. Current cloud computing makes everything a service that will finally be extended to every consumer.

Shang Bing, deputy minister of the MIIT, said, "China's current cloud computing has had some foundation; its application practices are unfolding; the infrastructure is improving; the industry chain has started to form."

**Cloud TV helps family life enter smart times**

Cloud TV is a new generation information terminal product aimed at the integration of the three networks/4C. It incorporates a computer, set top box, IPDVD, game console, digital family centre and home gateway. It fully embodies the development trend of digital information terminal products: "computer technologies as the platform, communication technology as the bond and consumer electronics as the product modality".

First generation cloud TV products were jointly developed by Hunan CATV and Peking University over a period of five years. Its main functions included: supporting 3D HD broadcasting, including Blue-ray technology; remote video call and conferencing; supporting PC functions such as text processing, email, online chat and web surfing; smart home and various value-added information services.

## The internet of things

### *Interpretation of key terms*

The internet of things is also called the sensor network, which is an extra large network combining all kinds of sensor equipment with the internet, such as a radio-frequency identification device, infra-red sensor, GPS and laser scanner. Its purpose is to let all things be sensed and controlled remotely, and connected with the existing network, to form a more intelligent production and life system. The internet of things is known as the third wave of world information industry after the computer and internet. It will also be the battleground in the new round of competition in the information industry.

In everyday language, the internet of things is an internet where all things are interconnected. It has two meanings: first, it's an extension of the internet, but its core and basis are still the internet; second, its clients are not restricted to individuals, but include everything.

Characteristics such as self-organisation, miniaturisation and perception of the outer world makes the internet of things a broad application space. It could be used to predict the possibility of flood, fire and other disasters, real-time monitoring of patients' health and control of factories' automatic production lines.

Experts predict that, within 10 years, the internet of things will be widely popular. It has extensive use, covering many areas such as intelligent transportation, environmental protection, government affairs, public security, housing and accommodation, intelligent firefighting, industrial monitoring, elderly care and individual health. It is estimated by experts that this technology will be developed into a high-tech market of more than Rmb1,000bn in 10 years.

## Knowledge link

### China's internet of things

- The Outline of the National Program for Long- and Medium-Term Scientific and Technological Development (2006-20) and the major special project of the "new generation wireless broadband mobile communication network" put the sensor network into a key research area. China stands at a similar starting point to the US, Germany and other Western countries when it comes to researching applications for the internet of things.
- Shanghai Institute of Microsystem and Information Technology, the Chinese Academy of Sciences was one of the earliest institutes to engage in the technological development of the internet of things. In November 2008, this institute signed a cooperation agreement with Wuxi city to build an engineering technology R&D centre of a high-tech micro-nano sensor network. Currently, the two sides are formulating an overall construction plan of the "Sensing China" centre, as well as industry planning to build the China Internet of Things Industry Academy, leading to the formulation of China's sensor network technology development and standards. The internet of things has been listed as one of the top six emerging industries cultivated and developed by Jiangsu province.

Wang Chen; released by Xinhua News Agency

**On 11 November 2011, citizens watch 3D TV at the 12th China (Tianjin) IT Expo**

- In March 2010, Premier Wen pointed out in the Work of the Government that China would vigorously cultivate strategic emerging industries, and specifically that the country should strengthen R&D and the application of the internet of things.
- On 20 October 2011, the Internet of Things China 2011 Conference and Exhibition was held in Wuxi. Experts believed that China's internet of things was still at a preliminary stage. The industry's core technology still hadn't been formulated, with a low application and demand level, and the commercial model was not yet clear enough. It was suggested that the industry transfer as soon as possible into the development stage of taking products, industry, demonstration, commercial use and the core market.
- On 29 March 2012, the launch ceremony for the construction of an exclusive operation platform of food and medicine, security data as the core network of China's internet of things was held in the city of Bozhou, Anhui province, marking the start of the core network system's construction of China's internet of things.

## *Sci-tech serves people's lives*

### Technology of the internet of things widely applied in people's lives

When consumers buy pork in a supermarket, they can check the where pig was cultivated, and how it was slaughtered and distributed. They can even find out what it was fed and which veterinary drugs were used during the cultivation process through a food safety enquiry machine.

Intelligent tracing to the source is an application of the internet of things' technology. The technology has already been applied in animal disease prevention and control, and the traceability of products' for quality and safety purposes. The internet of things' technology is gradually being applied in various areas to benefit people's lives.

The internet of things technology has been applied in areas such as medical treatment, public transport, food security and intelligent housing. The internet of things' monitoring system for the pharmaceutical cold chain can perform real-time monitoring of a vaccine's temperature during storage and transportation to ensure its safety; an intelligent public transport system can transmit video, service status and passenger flow data, thereby improving a city's public transport service quality by alleviating alleviate traffic jams and delivering real-time transport information to people's homes. It can also install the QR code identification system on mobile phones, enabling individual QR codes to be captured by a phone camera, and for online shopping payments to be made by phone.

# Chapter 3

# Biological medicine

Over the past 10 years, China has launched several major sci-tech projects in biological technology R&D, such as significant new drug inventions, the prevention and control of major infectious diseases and the cultivation of new GM varieties.

In the area of biological innovation, China has been supporting a series of national engineering research centres and laboratories on bio-information, stem cell research, biochips, antibody engineering, tissue engineering, new-generation vaccines and modern traditional Chinese medicines. It has also been promoting pharmaceutical and breeding industries to build an enterprise technology centre, so as to encourage enterprises to develop technological innovation in China's biological sector.

China has been implementing nine special industrialisation projects, including biological medicine, modern traditional Chinese medicine, green agricultural inputs and microbial production. China has also pushed ahead with the industrialisation of nearly 1,000 major achievements, such as an H1N1 influenza vaccine, GM insect-resistant cotton, polylactic acid and non-food bio-fuel. These achievements have given China a development advantage in some key areas of the bio-industry.

## Significant new medicine development

### China's development of significant new medicines is accelerated

During the 11$^{th}$ five-year plan, China accelerated the development of significant new medicines: 18 new drugs have been given new drug certificates; new drug registry applications have been submitted for 22 varieties and 137 new drugs are in the clinical research phase. Meanwhile, the transformation of drug varieties is also going smoothly. Thirty-six drug varieties are being supported in their technical transformation, playing an important role in addressing difficulties in the area of medicine use and high drug price.

Liu Haifeng, Xinhua News Agency

**On 29 October 2011, Han Jin, technical director for Shanghai Tasly Pharm, displays a new thrombolytic drug, "r-pro-UK", which is on a national Class A level. "r-pro-UK" was a high-tech bio-pharmaceutical achievement, with several independent IPRs through independent innovation by Chinese researchers**

China has also supported 15 comprehensive platforms for new drug R&D. Good momentum has been experienced in R&D capacity to create new drugs. The construction of a new drug incubation base has resulted in the investment of Rmb10bn from local government and enterprises, leading to indirect investment of more than Rmb20bn.

Meanwhile, China has accelerated independent innovation in medical products, with the emergence of a series of high-end products, such as the DBS (Deep Brain Stimulation) instrument for Parkinson's disease, the multi-slice spiral CT and the high-performance automatic biochemistry analyser. China has also achieved breakthroughs in the areas of electrical impedance tomography (EIT) and brain-computer interface technology. It has become a world leader in non-invasive continuous image monitoring.

China also continues to promote the heritage and innovation of traditional Chinese medicine (TCM). During the 11$^{th}$ five-year-plan, the government mobilised national efforts to undertake heritage research into the clinical experiences and thoughts of 210 elderly TCM doctors, and formed IT support systems.

**Small-molecule anti-cancer drug independently developed by China enters the market**

Icotinib, the molecular targeted small-molecule anti-cancer drug independently developed by China, entered the market in August 2011. This meant China was no longer dependant on imports of this type of drug.

The success of the icotinib project has made an important contribution to China's transformation from copying to innovation, from a big medicine country to a big medicine power. China is gradually forming a medicine innovation system. By combining production, academic work and research, the R&D of China's medicine is going through significant reform in innovation.

Icotinib is a new generation anti-cancer drug that targets the epidermal growth factor receptor subfamily of Tyrosine Kinases. It is China's first molecular targeted small-molecule anti-cancer drug with independent intellectual property rights. It is used to treat late stage non-small cell lung cancer.

**China achieves progress in using a nourishing kidney prescription to treat primary osteoporosis**

On 6 March 2012, the Assessment Meeting of the Nourishing Kidney Prescription in Treating Primary Osteoporosis was held at the China Academy of Chinese Medical Sciences. The experts group at the Bone Fracture Sub-association of the China Association of Chinese Medicine believes that the achievements made have reached an internationally advanced level.

Osteoporosis is a common disease severely threatening China's middle-aged and elderly people. In the past, there was a lack of treatment for osteoporosis, particularly primary osteoporosis. Through an epidemiological investigation of 6,447 high-risk people in Beijing, Shanghai, Fujian and other places, the research team worked on the theory that "the kidney dominates the bones" and explained the etiology and pathogenesis of osteoporosis, as well as the mechanism of the action of a nourishing kidney prescription in treating primary osteoporosis. The research team investigated signal transmission between osteoblasts and osteoclasts. On the basis of network pharmacology, the team explored the interaction between osteoporosis proteins, and the prediction of pathways underlying molecular complexes; using modern science and technology to find from the osteopractic single herb that the osteopractic total flavone is an active substance to promote the formation of bone cells, and show that it could promote the differentiation and proliferation of osteoblasts. During the preparation procedure for the osteopractic total flavone, part of the standard is formed for the SFDA "Macroporous Resin Absorption Technique", and relevant research is carried out on the clinical effectiveness, safety and action mechanism; the "Chinese Medicine Evidence-Based Clinical Practice Guidelines for Primary Osteoporosis" has been developed.

This research subject has won a domestic invention patent, new drug certificate and four international invention patents. After 20 years of R&D, nourishing kidney prescriptions and primary osteoporosis has been promoted and adopted at more than 520 medical institutions and more than 180 community health service stations across the country, showing good clinical effects in aspects such as significantly improving the osteoporosis patients' ostalgia, increasing bone density, alleviating lower limb myasthenia and clonus, and enhancing patients' quality of life.

**Breakthrough in the localisation of new anti-cancer drug**

Two new anti-tumour drugs, PEG-rhG-CSF (Pegylated Recombinant Human Granulocyte Colony Stimulating Factor for Injection) and Doxorubicin Hydrochloride Liposome Injection, independently developed by CSPC Pharmaceutical Group, entered the market on 17 March 2012 after being approved by the State Food and Drug Administration. This was a significant breakthrough for CSPC in the transformation from a crude drug to a high-end drug, and it broke the monopoly of foreign pharmaceutical enterprises in this type of product.

The two new approved drugs from CSPC belong to the National Significant New Medicine Development project. PEG-rhG-CSF was the first domestic drug to enhance white blood cells in the long-term. Doxorubicin Hydrochloride Liposome Injection was the first anti cancer liposome drug independently developed by CSPC, which is recommended by the National Comprehensive Cancer Network (NCCN) ovarian cancer guidebook as a frontline drug.

**China's first therapeutic TCM approved by the EU to be registered and marketed**

Diao Xinxuekang capsules, jointly developed by Chengdu Institute of Biology and Diao Group, were approved by the EU to be registered and marketed on 22 March 2012. This was a breakthrough for China's therapeutic TCMs with independent IPR to enter the main market of developed countries. It also became the first botanical drug outside the EU to gain market access.

Diao Xinxuekang capsules, created in 1988, are China's first modern TCM variety with a completely independent IPR. This was the culminations of eight years of work by Li Bogang and his team, who overcame serious technical difficulties to achieve large-scale industrial production of high-purity steroidal saponin. The capsules have been widely applied in clinical treatment in China over the past 20 years.

## The prevention and control of HIV-AIDS, hepatitis and other major infectious diseases

During the 11$^{th}$ five-year-plan, China invested about Rmb3.4bn in a project to prevent and control HIV-AIDS, hepatitis, and other major infectious diseases. A sci-tech research team of nearly 10,000 people was formed, and they made a series of technological breakthroughs and created a batch of products.

Phase I of the clinical trial has been concluded for the DNA/replicating recombinant vaccinia virus vaccine, and phase II of the clinical trial is under preparation. In those groups undergoing clinical trials, the death rate of children with HIV has been reduced to 38.8 per 100 in 2003 to 0.71 per 100 in 2010.

Since frontline treatment for HIV-infected adults started, side effects have been remarkably decreased. The cost is only one-twelfth that of imported medicines; a high-dose (60 micrograms) hepatitis B vaccine has been successfully developed, with more than 80% being cured as a result of a single injection. It can be used to control hepatitis B in a floating population, and there are hopes that new infection rates will be lowered.

Yi Ke, a therapeutic vaccine with an independent IPR by China has completed phase III clinical trials conforming to international regulations at the end of 2010.

China has successfully developed a series of tuberculosis diagnostic products, improving the detection rate of tuberculosis bacteria, and shortened testing time.

The country has made great progress in preventing and controlling HIV/AIDS and other infectious diseases; it has also rapidly advanced sci-tech to help lower fatality rates for some major infectious diseases.

Data shows that, with the interaction between different departments, more than 100,000 HIV/AIDS patients have received treatment in China, and the fatality rate has gone down from 16.1 per 100 a year in 2008 to 13.5 per 100 a year in 2009.

Through many years of effort, the surge in incidence of HIV/AIDS in China has been restrained and the goal of controlling the number of infected persons to within 1.5m was realised at the end of 2010.

In the area of tuberculosis prevention and control, according to the fifth national epidemiological sampling survey for tuberculosis in 2010, the TB morbidity rate in Chinese people aged 15 years and over was 66 per 100,000, down by more than 60% from the 169 per 100,000 figure in 2000.

At the same time, China has formed a networking technological security system and established a sci-tech supporting system for the prevention and control of major infectious diseases.

A monitoring platform for infectious diseases has promoted the transfer of monitoring and detection technology from the central government to the local level. A comprehensive demonstration area for prevention and control has realised a regional concentration of resources, deployed five comprehensive demonstration areas and 10 on-site research subjects of single diseases, and made significant progress in infrastructure construction. In an area containing around 20m people, important survey and e-archiving work has been carried out. An experiment effort for comprehensive prevention and control measures has been strengthened.

The major prevention and control project for infectious diseases improves the ability to tackle "traditional" and "emerging" infectious diseases.

In the past five years, we have seen the emergence of avian flu, H1N1 influenza, hand and foot and mouth disease, tick worm disease and the superbug. Thanks to the programme of the infectious disease detection technology platform, China has formed a network laboratory consisting of 14 leading labs, 77 cooperative labs and 240 hospitals for detecting infectious disease. This has made China better prepared to fight outbreaks of infectious diseases.

When the H1N1 influenza epidemic appeared in 2009, it was the ability to develop diagnostic reagents within 72 hours that gave China the ability to address the epidemic. As the first country to have developed an H1N1 influenza vaccine, China has played a key role in controlling a large-scale influenza epidemic. In view of the enhancement of China's ability in influenza detection and monitoring, as well as the recognition of its ability to tackle the influenza epidemic, the WHO appointed the Chinese National Influenza Centre as the world's fifth, and the developing countries' first, WHO Collaboration Centre for Reference and Research on Influenza.

In the spring of 2011, another piece of good news arrived. China became the first country in the world to verify that the fever with thrombocytopenia syndrome was an emerging infectious disease caused by a novel bunyavirus. This led to the discovery that the disease was transmitted by a tick.

The discovery of the novel Bunyavirus was another major breakthrough for China in etiology since the discovery of the SARS corona virus, and it has earned the country considerable international recognition.

## The cultivation of a new GM biological variety

- There is a technology that can transfer sections of a bacteria's DNA into cotton plants, causing the cotton bollworm to die after it eats the plant.

- There is a technology that can effectively control lepidopteran pests and secure an increased output of rice, and at the same time reduce the dosage of chemical pesticide by 80%.

- There is a technology that can increase milk protein by 10%, with important benefits such as improving immunity, promoting neurodevelopment, enhancing iron absorption and improving anaemia.

That is transgenic technology.

"The major breakthroughs in biological technology are nurturing new industrial revolutions. The development of China's biological industry has significantly promoted the R&D and industrialisation of GM varieties. This bears an important strategic meaning for strengthening China's core high-tech competitiveness and securing China's food security and sustainable agricultural development," said Xu Zhihong, a member of the Chinese Academy of Sciences.

The production of human lactoferrin from the milk of transgenic cows is a new GM variety that China has developed with an industrialised application.

Human lactoferrin is a kind of natural protein found in human breast milk that helps babies to grow and develop. It contains plenty of iron, improves the immunity and has anti-bacterial and anti-tumour effects. Currently, infant formulas in domestic and foreign markets mostly add lactoferrin that is purified from milk, but with a much lower biological activity level than human lactoferrin. Due to the fact that it comes from human breast milk, the sources of human lactoferrin are quite limited, and therefore hard to produce on a large scale by means of common technology. The human lactoferrin milk from transgenic cows contains 10% more protein than normal milk. Experiments also prove that it has special functions such as improving immunity, promoting iron absorption, improving anaemia and has anti-bacterial and anti-tumour effects.

This technology is going through GM biosafety tests. These were the first GM animals to enter productive experiments in China.

During the 11$^{th}$ five-year plan, the Outline of the National Programme for Long- and Medium-Term Scientific and Technological Development (2006-20) identified new GM varieties as the only special programme. In 2009, agricultural biological breeding was listed on the national development plan for strategic emerging industries.

"Transgenic technologies provide effective solutions for many problems in agricultural production, such as increasing crops' resistance ability, preventing and controlling pests, reducing the use of chemical pesticides and lowering environmental pollution and harm to other insects, humans and animals," said Li Jiayang, vice chairman of the Chinese Academy of Sciences.

Experiments carried out by scientific researchers led to the discovery of several genes with important application values. Based on this, they have developed new varieties of GM crops capable of disease and insect resistance, created new GM plant and animal materials, and enhanced China's R&D innovation ability of GM organisms, thus providing strong technological support for China's sustainable agricultural development.

Gene-modified cotton is one of China's key GM technological research results. Through biotechnology, it transfers segments of a bacterium's genetic material into the cotton plant to make it pest-resistant and causing cotton bollworms to die after eating it. The application of this pest-resistant GM cotton could lower production costs, reduce pesticide hazards, protect the environment and become an example of innovation and application for China's agricultural GM technology.

During the 11$^{th}$ five-year plan, China took the lead internationally in establishing a three-line hybrid cotton molecule system. The four new anti-pest cotton varieties developed have passed national examination, being cultivated in fields covering more than 2,670 sq km, and have generated economic and social benefits of around Rmb1.1bn.

The research team, consisting of Li Jiayang, Qian Qian and other researchers at the Rice Institute of the Chinese Academy of Agricultural Sciences, have made breakthroughs in enhancing rice production. They have successfully cloned the key multi-effective gene IPA1 determining the ideal rice plant type.

The ideal plant type is a core area in the current super-rice plant research, which improves rice yield by changing the rice plant's structural characteristics. The mutation of IPA1 would reduce the rice plant's tillering, increase the number of spikes and grains, boost grain weight, make the stalk sturdier, and enhance the lodging-resistant ability. Experiments show that, by putting the mutated gene into a conventional rice plant variety, yield can increase by around 10%.

## Stem cells

A stem cell is a kind of protocell with self renewing and multiple differentiation properties. It is an organism's cell of origin, with the protocell growing into various human body tissues and organs. Under certain conditions, it can develop into diverse functioning cells, tissues or organs, and therefore is called "a multi-purpose cell" by medical practitioners.

China was the first country to report the molecular marker determining the totipotency of stem cells, meaning they have the potential to create an entire organism. This was of great significance for maintaining the mechanism of stem cell pluripotency, by which stem cells have the potential to differentiate into almost any cell in the body. It also helped to promote the screening and isolation of stem cells, and establishing and maintaining therapeutic cloning, inducing transplanted cells.

China first used a self assembly method to make Nano-HA/Collagen bone rehabilitation material that is quite similar to autogenous bone, and is quite compatible with the human body. It also has the advantage of no immune rejection, and is a new generation of bone implantation material.

**Chinese scientists successfully cultivated the first domestic GM macaque in Kunming, Yunnan province. This marked an elevation of the work of Chinese scientists in the area of non-human primate GM animal research. Before that, only American and Japanese scientists had successfully acquired the GM monkey model. Image above: The GM macaque is no different to the ordinary macaque in appearance. Image below: under special light, the body of the GM macaque (left) looks green. This is due to a green fluorescent protein in its body**

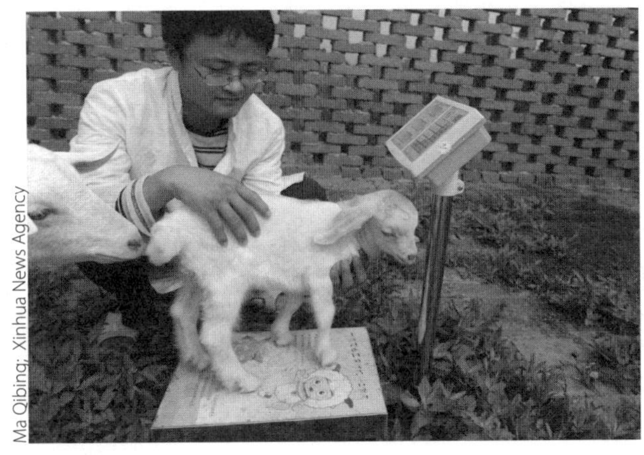

Ma Qibing; Xinhua News Agency

**Anhui's first "second generation" GM clone goat completes its first month of life. On 13 June 2011, at a breeding base in Feidong county, Hefei city, workers measure the weight of the GM kid**

The "active spinal cord reconstruction tubes" independently developed by China can effectively promote recovery from spinal cord injury. It is the first internationally approved product of the spinal cord injury restoration project that has entered clinical research.

## Knowledge link

### Medical uses of stem cells

Stem cells have great potential in clinical medical care. Stem cell treatment designed to replace or repair injured tissues involves transmitting stem cells from either autograft or xenograft into diseased tissues by way of blood vessel infusion or local injection.

The theory of disease is that when cells or tissues are denatured, die or their functions are weakened or lost, then disease follows. Stem cells have the special biological ability to duplicate and differentiate into diverse kinds of cells. Therefore, replacing or repairing patients' injured cells with stem cells could restore the function of the cells and tissues and thus reach the goal of disease treatment.

Theoretically, stem cell technology could cure various diseases of many systems such as the nervous system and the circulatory system. Furthermore, it has many advantages over the traditional treatments:

- Safe: low, or no toxicity;
- Can be used when the specific pathogenic mechanism is not yet fully known;
- Adequate resources for treatment materials;
- Vast treatment scope;
- The best immune treatment and genetic therapy carrier;
- Diseases that were incurable using traditional therapy might be treated in future.

Stem cell treatment could be effective in treating those diseases where current medical methods are unsatisfactory, such as cancer, myocardial necrosis diseases, immunological diseases, leukaemia, liver disease, kidney disease, diabetes, senile dementia, Parkinson's disease, spinal cord injuries and skin burns. Combined with gene treatment, stem cells could cure many hereditary diseases.

Currently, stem cell treatment has six common methods: transradial intervention, local seeding, vena approach, lumber puncture, stereotactic encephalic stem cell transplantation and CT-guided spinal stem cell transplantation.

## Chinese scientists find solutions to molecular barriers of the somatic cell "changing" into a pluripotent stem cell

According to a report by Xinhua News Agency on 18 November 2011, Chinese scientists have found a solution to the molecular barriers of vitamin C promoting the transformation of a somatic cell into an induced pluripotent stem cell. This has laid the foundation for creating the formation mechanism of an induced pluripotent stem cell.

An induced stem cell happens when, under the induction of an adventitious agent, a somatic cell "changes" into a pluripotent stem cell with the same characteristics as an embryonic stem cell outside the body. It is of great significance in tissue and organ transplantation and gene therapy. It is also helpful to generate an important influence in new drug screening and development, gene discovery and toxicity assessment. Although there are many possible applications for induced pluripotent stem cells, some problems have been evident to scientists for quite a long time, such as the unknown induction mechanism and low induction efficiency.

In 2009, Pei Duanqing and some other researchers found that vitamin C could greatly improve the efficiency of a somatic cell transforming into an induced pluripotent stem cell. The team conducted a lot of basic research and found that vitamin C can improve the transformation efficiency of a somatic stem cell through the lowering of the molecular barrier by a special enzyme.

After much screening, scientists found an enzyme that could significantly improve the cell reprogramming efficiency. This enzyme and vitamin C could both accelerate the growth of the adult cell,

Experiment results showed that, when an untreated somatic cell-fibroblast goes down to the sixth generation, they are almost too old to transform into a pluripotent stem cell. But after the introduction of this enzyme and the addition of vitamin C in the culture medium, the fibroblast still hasn't shown an aging phenotype when going down to the sixth or even 12th generation outside the body. Under this condition, the fibroblast could keep the same "transformation" potential as the original cell, and maintain reprogramming efficiency.

Dr Villiger, a cell biologist at Stanford University, said, "This research result explains the synergistic effect between the protein and vitamin C. It could open the "turned-off gene" that is necessary to complete reprogramming, and therefore push the reprogramming ahead. It is a milestone discovery for people to understand the cell reprogramming mechanism on a molecular level, and of deep significance for cell and regenerative medicine."

### Stem cell technology might tackle diabetes and thalassemia

According to a report by Xinhua News Agency on 20 December 2011, thalassemia, Parkinson's disease, and diabetes are incurable diseases that could be cured in future by a medical technology called "induced pluripotent stem cell". China has an advantage in stem cell research in aspects such as etiology and diversity of diseases. The country is expected to be at the forefront of future discoveries in this area.

The pluripotent stem cell is a kind of stem cell that is capable of maintaining self-renewal and multi-potential differentiation. Its main characteristic is that it can differentiate the various cells that the human body needs. The new differentiated cell would not be rejected after being reintroduced into the human body.

## Genetic engineering

### China's leads the world in genomic sequencing

China has built up the world's largest genome sequencing platform, ranking first in the world in sequencing ability, and achieving leapfrog development on genomic technology. This has greatly enhanced people's ability to learn about nature and has provided innovative supporting technology to safeguard people's health and increase agricultural and industrial production efficacy.

### Test chip for hereditary deafness gene created by China for possible clinical use

Each year, nearly 30,000 babies are born with congenital deafness, and 60% of these cases are caused by a hereditary factor. The application of gene-detecting technology in clinical practice could lead to significant advance in this area.

According to a report by Xinhua News Agency on 9 February 2011, the gene detection chip for rapidly diagnosing hereditary deafness, independently developed by The National Engineering Research Centre for Beijing Biochip Technology and CapitalBio Corporation, was the first to receive a medical device certificate issued by The State Food and Drug Administration. It is expected to go into large hospitals across the country in the coming years.

The gene-detection chip embeds the nucleic acid segment of a known gene sequence as a bioprobe onto a fingernail-sized slide or silicon wafer. It then sends out a signal by reaction with a sample, and uses computer technology to collect signal data and analyse the sample's gene mutation to diagnose a hereditary disease.

The gene detection chip is highly efficient and relatively cheap. It is conducive to improving the lack of early diagnosis of hereditary deafness in many regions in China. Cheng Jing, director of the National Engineering Research Centre for Beijing Biochip Technology and a member of the Chinese Academy of Engineering, said that this chip could work before childbirth to give new parents hereditary information about their baby and take measures to lower the probability of newborns suffering from hereditary deafness.

A considerable proportion of deaf people in China are gene mutation carriers. When they were born, they were not deaf. They went deaf because of the injection of aminoglycosides, such as streptomycin and gentamicin.

"Many doctors know that this kind of medicine could lead to gene mutation and then cause deafness," said Cheng Jing. "But due to shortcomings in gene detection, we can't differentiate [gene mutation carriers] from normal people. Our chips could detect the gene mutation position of 90% of instances of drug poisoning deafness, and therefore greatly reduce the tragedy of "one injection causes deafness"," said Cheng Jing.

CapitalBio Corporation developed the world's first gene detection chip for hereditary deafness in 2007, and received a national medical device certificate in 2009. Currently, the chip has carried out gene detection for tens of thousands of people, of whom more than 30% were found to be carrying the ear gene mutation that can cause gene mutation deafness.

**World's first whole genome breeding chip of rice is born in China**

According to a report by Xinhua News Agency on 3 May 2012, China National Seed Group Corporation, a wholly owned subsidiary of Sinochem, cooperated with Huazhong Agricultural University and Peking University to jointly develop the world's first whole genome breeding chip of paddy rice. It is designed to improve the accuracy of identifying genuine seeds and is conducive to improve seed-breeding efficiency and eradicating the damage caused by fake seeds.

Currently, when judging the authenticity of rice seeds, industry practitioners usually adopt the internationally-recommended 24 SSR (simple sequence repeat) test results. The new chip could take advantage of more than 5,000 loci across the whole genome to come up with more accurate and reliable test results.

This gene chip also has advantages in the background analysis of seed-breeding materials, and could judge the background of the breeding mediate material to accurately control and breed seeds. Therefore, it can help seed-breeding experts to significantly improve rice efficiency.

Traditional seed breeding has a long and unpredictable cycle, and is reliant on experts performing visual screening. With this new tool, large-scale and precise screening could be moved from the fields to the laboratory, removing more than 95% of single plants, with the rest being planted in the fields. In this way, field workload is significantly reduced. The former seed-breeding cycle of eight to 10 years could be reduced to just three to five years.

## Vaccine R&D

China has always paid close attention to the production and R&D of vaccines. During the 11$^{th}$ five-year plan, China launched several key projects to vigorously advance R&D into vaccines.

New vaccines such as the orally immunised recombinant Hp vaccine, Hepatitis E vaccine and A/H1N1 flu vaccine have helped accelerate China's emergence as a world vaccine power.

**Assistance: key projects boost vaccine R&D**

During the 11$^{th}$ five-year plan, in order to boost the vaccine industry, the Ministry of Science and Technology decided to include vaccine R&D and industrialisation in special national sci-tech projects such as the "863" plan (The National High Technology Research and Development Programme of China), the science and technology support programme, the Significant New Medicine Development, and the prevention and control of viral hepatitis, HIV/AIDS and other major infectious diseases.

Driven by several sci-tech plans, the innovative vaccine products independently developed by China for the prevention and control of major infectious diseases become the brightest "stars":

- The DNA-Tiantan Vaccinia Combined AIDS vaccine jointly developed by the National Centre for AIDS/STD Control and Prevention, the China CDC (The Chinese Centre for Disease Control and Prevention) and the National Vaccine & Serum Institute received approval to enter clinical trials at the end of 2006;

- the inactivated poliovirus vaccine prepared with the Sabin strain and developed by the Institute of Medical Biology, the Chinese Academy of Medical Sciences obtained approval to enter clinical trials in May 2007;

- the Therapeutic HBV-DNA vaccine jointly developed by Guangzhou BaiYunShan BaiDi Biotechnology, the $458^{th}$ Hospital of the PLA and Guangzhou Pharmaceutical obtained approval to enter clinical trials in August 2007;

- the HBV therapeutic vaccine developed by the Third Military Medical University obtained approval to enter Phase II clinical trials in August 2009;

- on 25 May 2011, the Sinovac Biotech announced that the EV71 (enterovirus type 71) inactivated vaccine, a preventive vaccine it had independently developed for hand, foot and mouth disease passed Phase I clinical trials.

**Innovation: from catching-up to leading**

Unremitting and independent innovation allows China to lead the world in the R&D of some vaccine varieties.

- In March 2009, the State Food and Drug Administration issued a new drug certificate to the orally immunised recombinant Hp vaccine, whose IPR is completely owned by China.

- When the A/H1N1 flu broke out, China's independent innovation was quick to react to this menacing problem.

"China becomes the first country in the world capable of applying the A/H1N1 flu vaccine." This major news was announced by Chen Zhu, China's health minister, on 8 September 2009, giving people hope that the epidemic could be brought under control. It is by means of the Alternative Method for Determi-

Wang Zhongwei Xinhua News Agency

**China successfully developed the world's first Hp vaccine. This photo shows a researcher checking the vaccine**

nation of Haemagglutinin Content and the interim standard reagent that China does not have to wait for the offer of a standardised reagent from the World Health Organisation to significantly shorten the time of clinical trials, and therefore move ahead of other countries in the R&D of an A/H1N1 flu vaccine.

- On 11 January 2012, China's Ministry of Science and Technology held a press conference in Beijing to announce that the Recombinant hepatitis E vaccine jointly developed by Xiamen University and Yangshengtang-Wantai Co had received a national Class A new drug certificate and production code, becoming the world's first vaccine used for preventing Hepatitis E. This was a milestone for China's original innovation in vaccines, and a great contribution by Chinese scientists towards the prevention and control of hepatitis across the globe.

The progress of research into hepatitis E has always been closely followed by the international pharmaceutical sector. The team issued 26 academic theses in world-famous academic journals, and they were invited many times to report on research progress in international academic and vaccine industry conferences.

The hepatitis E vaccine's access to the market was a milestone in the global prevention and control of hepatitis. The team will introduce the vaccine to high-risk groups in China as soon as possible. It is also cooperating with the WHO and philanthropic organisations to try to bring benefits to people in other countries and regions that have a high incidence of hepatitis E.

# Chapter 4

# Advanced manufacturing and new materials

Over the past decade, China has made great achievements in the equipment manufacturing industry. In terms of scale, China is now ranked in the leading position. There has been a significant improvement in the indigenous level of major technical equipment. Reflecting the status of equipment manufacturing as a pillar industry, a batch of key products has reached an internationally advanced level; a group of leading equipment manufacturing enterprises has come to the forefront; a group of industrial investment zones is shaping up quickly; the market share and technical level of several products have improved in world terms. During the 12th five-year plan, the restructuring and transformation of development patterns will become a top priority for the equipment manufacturing industry to realise scientific development. We will grasp the core technology of the principal products, own the IPR to a range of key products and well known brands, develop strategic emerging industry equipment, produce more high-end equipment to the extent that it accounts for more than 15% of the total, and secure the overall needs of the national economy and national defence sector.

After much hard work over a period of many years, China's new materials industry has been developed from scratch. The industry has grown and developed with remarkable achievements in areas such as system construction, industry scale and technical progress.

New materials are being used in an ever wider range of economic areas, and they can be found throughout the production process from R&D and design, to production and application.

The breadth of the new materials industry continues to grow. China is now one of the world leaders in rare earth materials, advanced energy storage materials, photovoltaic materials, organic silicon, super-hard materials, special stainless steel, glass fibre and its composite material rank prominently in world terms.

Major breakthroughs have been achieved for some key technologies. Production technologies developed by China have reached the international level, including beryllium tantalum niobium alloy, alloy amorphous alloys, high magnetic induction grain-oriented silicon steel sheets, diphenylmethane diisocyanate (DPMDI), super-hard materials, m-aramid fibres and super conducting material. New varieties and enhanced functionality are increasing in high-end structural metallic materials, new type inorganic non-metallic materials and high-performance composite materials, while the country is becoming more self-sufficient in advanced polymeric materials and special metallic functional materials.

## High-grade CNC machine tool and basic manufacturing equipment

### *Interpretation of key terms*

High-grade computer numerical control (CNC) machine tools and basic manufacturing equipment is one of 16 major national special programmes. This special programme mainly develops basic equipment manufacturing including high speed, precise and combined numerical control machine tool and technical processes such as hot working and surface preparation.

This special programme has been designed and implemented by the Ministry of Industry and Information Technology. Following the guiding principle of "tracking national advanced technologies, strengthening the basis, leapfrog development, integrated innovation, grasping core technology, enhancing innovation ability, encouraging use and driving demand", focusing on key areas such as aerospace, marine transportation, vehicle and power-generating equipment, key tasks in seven aspects are deployed: main engine, numerical control systems, functional and key unit development, generic technology development, innovation platform, user's technological application test and application demonstration project construction.

By 2020, we will have enhanced the independent development ability of high-grade CNC machine tool and basic manufacturing equipment products, raising the technology level to that of internationally advanced products. A complete functional unit will be established, along with a technical innovation system with enterprises as the backbone and an industrial university research combination. A high quality R&D team will be established and a number of innovative technologies and products will be developed that carve out a significant market share. The aim is to become 80% self-reliant on high grade CNC machine tool and basic manufacturing equipment that are needed for the aerospace, marine transportation, vehicle and power generating equipment sectors; develop several original technologies and products and make a breakthrough in self innovation, with a significant increase on market share.

### *Result highlights*

During the 11th five-year plan, the main achievements of the high grade CNC machine tools and basic manufacturing equipment special programme are:

Development of major equipment reached an internationally advanced level. Focusing on emergency needs from four major areas including aerospace, marine transportation, vehicle manufacturing and power generation equipment, China has mastered a series of key technologies such as super-heavy machine tool design and manufacturing technology, super-large vertical and horizontal special drilling machine design and manufacturing technology, ultra-wide and ultra-long work piece processing and developing a number of internationally advanced manufacturing equipment such as a numerical control "heavy gate bridge", "five axis turning and milling multiply cutting machine" and a large, fast, high, efficient and fully automated "impact forging line".

The performance of middle- and top-grade numerical control machine tools has been enhanced and the initial domestic demands met. China has developed products with a wide range of market applications, such as a high speed, a precise processing centre, numerical control machine tools and a five-axis turning and milling processing machine tool. It has also made key technological breakthroughs in terms of reliability, digital design, overall performance rating and dynamic compensation, and effectively enhanced the technological level and market share of middle- and top-grade numerical control machine tools.

Wang Xin;Xinhua News Agency

**On 16 April 2012, a manufacturer (first from left) at the 7th China CNC Machine Tool Fair is introducing CNC gear-hobbing machine tool products to clients**

Ren Yong, Xinhua News Agency

**A wheel automatic flexible machine unit exhibited at the 12th China International Machine Show by Shenyang Machine Tool on 11 April 2011. Robots finished the automatic carrying in work piece processing for this set of automatic machine units, realising a high degree of automation**

R&D of the numerical control system, the functional unit and the cutter are going smoothly, with the emergence of a generic technological achievement. In particular, remarkable breakthroughs have been achieved in key technology of the numerical control system and have been applied to domestic machine tools. China has successfully developed the numerical control system, matching the five axis turning and milling processing machine tool to an international standard. Partial technical breakthroughs have also been realised in functional units such as the ball bearing screw rod and the straight line lead rail, the numerical control rotary worktable and the double angular numerical control universal milling head. China has developed a number of precise, complex and hard alloy cutters and applied them in key enterprises, with some of them replacing imported products.

## Ultra-large-scale integrated circuit manufacturing equipment and the complete set of technologies

### *Special achievements*

Ultra-large scale integrated circuit manufacturing equipment and the complete set of technologies is one of the 16 major national programmes. This special programme aims to develop key manufacturing equipment of the integrated circuit, to own the IPR for complete sets of advanced technology and other relevant new material technology, thereby breaking the reliance on imports of high-end integrated circuit manufacturing equipment and technology and leading to the technical enhancement and restructuring of relevant industries.

Since the implementation of ultra-large-scale integrated circuit manufacturing equipment and the complete set of technology special programme, China has mastered a number of core technologies that restricted industrial development in complete equipment, packaged technology and key materials, leading to the enhancement of China's competitiveness in the manufacture of integrated circuits.

In terms of packaged technology, this special programme has finished the complete R&D for 65nm packaged product technology and has started batch production, making China reach an internationally advanced level for the first time. China has also achieved a breakthrough in 12-inch key complete machinery products and components, challenging the established monopoly of foreign companies. The 12-inch 65nm dielectric etcher has passed process validation and received a bulk order for 20 units, both in and outside China.

These achievements have greatly narrowed the technological gap between China and developed countries, stimulating many domestic integrated circuit design enterprises and electronic complete machine enterprises to enhance their overall competitiveness.

### *Extended reading*

#### AMEC etcher: the wisdom in China for the semiconductor industry to go global

"A handy tool makes a handy man". The support of equipment manufacturing is essential for the development of the semiconductor industry. For a long time, foreign companies have dominated the global market for high-end semiconductor equipment, due to the high technical threshold. This was particularly true for super high-precision etching.

With support from major national sci-tech programmes, the Shanghai-based Advanced Micro Fabrication Equipment (AMEC) has successfully developed a 65-40nm dielectric etcher and entered the international market. This changed the competitive situation in the world's key etching equipment market and enhanced China's international competitiveness in the semiconductor industry.

What is a 65-40nm dielectric etcher and how does it affect people's lives? Dr Yin Zhiyao, chairman and CEO of AMEC, said, "If we describe the first Industrial Revolution as a revolution of machines replacing people's hands, in which people invented key devices such as machine tools, milling machines, facing machines and forged a traditional macro processing industry, then we may say that the second Industrial Revolution started 60 years ago in Silicon Valley in the US. This was a revolution of computers replacing human brains, in which the core technology was microscopic material manufacturing and microscopic processing. The etching equipment that we develop is just like the machine tools, milling machines and facing machines in microscopic processing.

"In contrast to macro processing, microscopic processing involves measurements of several thousandths of the width of a simple hair follicle. One plasma etching equipment can process dozens of trillions of microscopic structures a year. And the processing precision, uniformity and repeatability must be accurate to the degree of several tens of thousandths of a hair follicle," said Dr Yin.

In daily life, almost all modern household appliances require chips. The production technology of a chip the size of a fingernail involves hundreds of procedures, needing 10 types of microscopic processing equipment, of which a lithography machine, an etcher and a chemical film machine are the three most important.

The tiny chip is segmented from a bigger silicon wafer. The processing object of a 65-40nm dielectric etcher manufactured by AMEC is a saucer-shaped wafer.

The processing of a wafer is like constructing a 12-storey microscopic building, microscopic expressway or microscopic overpass on the basis of the round saucer. "First, put a layer of chemical or metallic film on the round saucer, then a layer of photoresist," explained Dr Yin. "Use a lithography machine for exposure and photo development so as to expose the surface to processing, and then use an etcher to cut and polish. At present, our etching equipment can process the microscopic structure at a minimum of 65-40nm. In the latest technical development, it can etch a microscopic structure at 32-28nm."

Due to its innovative design, an AMEC semiconductor etcher can process two wafers at the same time, while foreign high-end etching equipment can only process a single slice wafer. Because AMEC etching equipment has a strong chip processing performance, small floor space, high output efficiency and low cost, it is widely favoured by international semiconductor factories in competition with the most internationally advanced semiconductor chip factories.

After overcoming the initial high barriers, AMEC etching equipment has successfully entered the international market, going into the production lines of eight overseas clients in Asia and having already processed more than 1.5m pieces of wafer. Many process validations prove that AMEC products not only reach the level of similar foreign products in terms of various performance factors, but also have a chip output efficiency that is 30% higher than the same type of foreign products with the same equipment investment.

After careful evaluation, many Asian chip factories have chosen the etching equipment produced by AMEC. The world's leading chip manufacturers have also placed bulk orders, meaning that AMEC will soon reach an annual output value of more than US$100m.

The plasma etcher developed by AMEC has won more than 10 prestigious prizes in China, the US, Europe and Singapore, including the 2009 Best Product named by Semiconductor International, a leading US-based journal.

## Intelligent robots

### Low-altitude flying robot, Snow Goose

In China's 26$^{th}$ Antarctic science investigation activity in 2010, the polar region low-altitude flying robot Snow Goose carried several types of scientific research loads and conducted a low-altitude auto flight test in the severe Antarctic environment, including an observation of large-scale sea ice, and in the process gained many valuable polar sea ice images and data.

### Versatile intelligent robot

**On 23 May 2012, at the 15$^{th}$ China Beijing International High Tech Expo, a child and parent robot are demonstrating interactive dialogue**

Luo Xiaoguang, Xinhua News Agency

Li Xiang ,Xinhua News Agency

On 24 June 2011, during the Robots World Tour, several students watch a humanoid robot performing pas de deux

**China develops wearable robots to help disabled and the elderly**

According to a report by Xinhua News Agency on 3 April 2012, the demonstration platform of wearable robots to help the disabled and the elderly, which was jointly developed by Hefei Intelligent Mechanical Institute, the Solid Physics Institute of the Chinese Academy of Sciences and Anhui Institute of Traditional Chinese Medicine, had recently passed site acceptance by an expert panel.

It is a key project of advanced manufacturing in the national 863 Program, which intends to design a set of wearable robots to help the disabled and the elderly with certain motor functions and the disabled with complete upper or lower limbs. It can be worn on the arm, waist or leg to provide users with the ability to reach objects or help them walk.

Overseas-developed lower extremity exoskeletons, also known as wearable power assistance robots, can ascertain a person's motion intention by sensing myoelectricity and electroencephalogram (EEG) changes. Its electrodes must adjoin muscles or the head, making the application inconvenient, and easily influenced by environmental, personal and other external factors.

The project team developed a new exoskeleton power-assisted robot that could judge a person's movement intention by sensing the change of muscle force in the upper limb and leg and provide corresponding impetus. It has made a breakthrough in the key technique of power-assisted hip joints and in the area of predicting falling, using a smart and simple structure that is easy to use and can be worn on the outside of a garment.

The demonstration platform will be able to help disabled or elderly people live an independent life and work autonomously, thereby alleviating the workload of families and social services. The research achievements could be applied in a range of areas, such as heavy labour work, medical treatment, manufacturing and entertainment.

**China makes breakthroughs in key components of humanoid robot**

According to a report by Xinhua News Agency on 17 April 2012, the humanoid foot sensor system developed by CAS Hefei Institute of Physical Science had recently passed an experts' acceptance check by the Ministry of Science and Technology.

The system's ability to integrate sensing and identification of the complex ground environment was a breakthrough for China and meant the country was no longer reliant on imported technologies.

A humanoid robot must guarantee a steady and reliable two-footed walking pattern in order to be effective. Wu Zhongcheng, a researcher in charge of this research, says that key technologies have been developed to acquire information and integrate it in the humanoid foot. Researchers then established an integrated foot perception and dynamic complex ground reaction simulation system, allowing real-time synchronised access that takes into account terrain, foot tilt and walking speed. The system has been tested on a BHR-2 robot platform at Beijing University of Technology Institute of the Intelligent Robot. Experimental results show that the system is well suited to practical applications.

This achievement has enhanced China's international competitiveness in humanoid robot research, promoted its research in humanoid robot theory and bionic control technology, and provides key technologies and reference for intelligent artificial limbs in terms of environmental adaptation, sensing and feedback control.

## LED semiconductor lighting

### *Interpretation of key terms*

LED stands for light emitting diode, which is a kind of semiconductor light emitting device. It uses a semiconductor chip as light emitting material and releases photons through a current carrier, which emits excess energy after composite effects in a semiconductor, giving out red, yellow, blue, green, cyan, orange, purple and white light.

LED is called the fourth generation, or green, illumination source, with benefits that include energy saving, environmental protection, long service life and small size. It can be widely used in areas such as display lighting, decoration, backlight and urban illumination.

Pei Xin, Xinhua News Agency

On 4 December 2009, constructors install LED energy-saving lamps for the Sun Valley Expo Axis at Shanghai World Expo. More than 80% of nightscape lighting at Shanghai World Expo Park used LED

Wang Zhen, Xinhua News Agency

On 29 February 2012, at the China Lighting Expo, visitors look at a new LED road lamp

## *Extended reading*

### Semiconductor lighting: technological application leads the world

At the China Central Television Spring Festival Evening Gala for the lunar New Year of the Dragon, LCD made an appearance on stage, demonstrating the rapid development of China's semiconductor lighting industry. In 2011, the size of China's semiconductor lighting industry had reached Rmb156bn, with a yearly growth of 30% and about 5,000 enterprises, making China the world's fastest growth region for this sector.

The reason for the rapid emergence of China's semiconductor lighting industry in recent years was due to the foresight of the Ministry of Science and Technology. The ministry was quick to promote development and cultivate the market and main market players. It also adopted a new model of setting up a technical innovation alliance, thereby promoting innovation and cooperation in universities.

**R&D and application nearly synchronised with international community**

Developed countries such as the US and Japan have been promoting R&D in semiconductor lighting technology as part of their industrial national strategies in the past decade. However, nine years ago, the upstream industry of China's semiconductor lighting technology sector hardly existed, while the downstream application had just started and all power chips were imported. The scale of the enterprises was too small, and they lacked important technology; overall sales revenue for industry was less than Rmb10bn.

To seize the first opportunity that existed here, from the 10th five-year plan onwards the Ministry of Science and Technology has supported the development of the semiconductor lighting industry.

The development of China's semiconductor lighting technology and industry has begun to accelerate with the support of the National Key Technologies R&D Programme, the National 863 Project, the sci-tech support project and the policies of individual departments.

During the 11$^{th}$ five-year plan, the efficiency of China's industrialised big power LED chips has reached beyond 100 lumens per watt, matching the level of mainstream international products. China now independently manufactures chips and epitaxial slices. The proportion of domestic chips made in China is increasing year by year, reaching 68% in 2011.

In addition, China has achieved significant progress in aspects such as the silicon substrate LED chips' industrialisation, low wavelength ultraviolet LED, MOCVD major equipment, the power typed white light LED packaging level. Statistics show that, in recent years, the number of Chinese LED patent applications has been growing by more than 30% a year. China has become a centre of global LED patent application.

**"Ten Cities, Ten Thousand Lights" illuminates emerging lighting industry**

To support the technological innovation and development of the semiconductor lighting industry, the Chinese government has made a series of policies and plans. In April 2009, the Ministry of Science and Technology launched the "Ten Cities, Ten Thousand Lights" semiconductor lighting application pilot project and approved 37 pilot cities for the plan, in two batches. In October 2010, the State Council's decision to speed up the cultivation and development of strategic emerging industries confirmed that semiconductor lighting was to feature in the key development direction of the new materials industry.

Under the "Ten Cities, Ten Thousand Lights" project, local governments in the pilot cities are trying to cultivate and develop the semiconductor lighting sector so that it can be an engine for adjusting the local industrial structure.

Many traditional lighting enterprises are entering the semiconductor lighting sector, opening the door for industrial transformation and upgrading. Ningbo is a centre for China's lighting industry. In 2010, among the more than 4,000 traditional lighting enterprises in the city, nearly one-third of them had switched to the semiconductor lighting industry, or were planning to do so. Some listed enterprises in traditional lighting also invested a large amount of capital into LED research and development. NVC Lighting, headquartered in Huizhou, Guangdong province, set up an R&D centre in Shanghai in 2010. The sparks ignited by Ten Cities, Ten Thousand Lights are now spreading. At the end of 2011, the Ministry of Science and Technology and the Ministry of Housing and Urban and Rural Development jointly held a meeting to discuss the project. Data released at the conference showed that more than 2,000 demonstration projects had been implemented in 37 pilot cities, with more than 4.2m LED light fittings installed, saving more than 400m kWh of energy. The pilot cities are home to nearly 4,500 semiconductor lighting enterprises, with a total output value of around Rmb140bn.

**First national key laboratory cultivated based on industry alliance**

On the last day of 2011, in a meeting room of the Semiconductor Institute of the Chinese Academy of Sciences (CAS), domestic and foreign experts on semiconductor technology and industry met to plan a structure for the National Key Laboratory for the Semiconductor Lighting Joint Innovation.

This lab is unusual. It is not headed by enterprises or sci-tech institutes or universities. Instead, it is led by the National Semiconductor Lighting Project R&D and Industry Alliance between university research departments and industry. It is the first lab of this type in China.

Since the launch of the National Semiconductor Lighting Project, China's semiconductor lighting has been developing rapidly. However, given the fierce international competition it still faces many challenges: the small scale of the enterprises, insufficient innovation ability, scattered R&D resources and a lack of a generic technology R&D platforms.

In 2004, backed by the Ministry of Science and Technology and other departments, 46 domestic backbone enterprises, universities and scientific research institutions jointly set up the National Semiconductor Lighting Project R&D and Industry Alliance.

With vigorous support from central and local governments, as well as the active organisation and coordination of the alliance, China's LED enterprises are thriving, with more than 10 of them successfully going public. The alliance itself also continues to develop. Currently, it has 293 member units, including the upstream and downstream elements of the industry chain, well-known enterprise and research institutions from Hong Kong, Taiwan and overseas, covering 40% of all LED enterprises and institutions in China.

**The International Semiconductor Lighting Alliance**

To transfer the huge market potential of China's semiconductor lighting industry into industrial advantage, on 16 October 2010 the National Semiconductor Lighting Project R&D and Industry Alliance in cooperation with semiconductor lighting industry organisations in the US, Australia, New Zealand, the Netherlands, South Korea, India and Taiwan region set up the International Semiconductor Lighting Alliance (ISA).

At a board of directors meeting, Wu Ling, secretary general of the National Semiconductor Lighting Project R&D and Industry Alliance, was elected as the first chairman of ISA. The secretariat of ISA was established in Beijing. The ISA is the world's first non-profit NGO for promoting the development of the semiconductor lighting industry. Its members cover the vast majority of leading enterprises in the industry, with rule-making power on semiconductor lighting. Until now, few organisations of this kind have been located in China.

"Guiding the setting of relevant international standards is one of the ultimate goals for industrial development and could fully embody the industry's discourse power in the world," said Yue Ruisheng, secretary general of ISA. He believed that, in terms of making semiconductor lighting applications, China and the West are at a similar starting point. China has a huge number of enterprises, sufficient production capacity and a wide market, which can hasten the introduction of many industry regulations and standards, so as to strengthen the influence of Chinese standards at an international level. (By Tang Ting) ("Science and Technology Daily", Page 9, 12 March 2012)

## New-generation recycling steel process technology

New-generation recycling steel process technology was first put forward by Chinese metallurgy experts in 2005 and involves integrating steel product manufacturing, high-efficient energy conversion and the absorption of large-scale social waste. During the 11th five-year plan, China took the relocation of Beijing-based Shougang Group as an opportunity to build a world-class, large-scale steel enterprise of Caofeidian Shougang Jingtang Steel Corporation, making an important contribution to the restructuring and upgrading of its steel industry.

Yang Shiyao, Xinhua News Agency

**On 6 April 2012, a ship docks at a quay in Caofeidian Harbour district to unload goods**

New-generation recycling steel process technology played an important supporting role in the construction of Caofeidian Shougang Jingtang Steel. This project led the development of China's steel process technology, and was the first design to follow the circular economy pattern and the first to use a natural deep water seaport in its construction. It was also the first in China to adopt a 5,500 cubic metre ultra-large blast furnace and the first to employ seawater desalination technology.

## Nanometer and superconducting materials

### China's first multi-functional nanometer photo catalytic air cleaner is successfully developed

According to a report by Xinhuanet.com on 15 June 2009, China's first multi-functional nanometer photo catalytic air cleaner with independent IPR was successfully developed at the National Engineering Research Centre of Ultrafine Powder in Shanghai.

This air cleaner has three disinfectant functions, photo catalytic bactericidal activity, ultraviolet sterilisation and adopts a carrier of high surface area and high absorption to carry the nano titanium dioxide photo catalyst web, which can release the coordination effect of high absorption and photo catalyst, realise the timely elimination of germs and the decomposition of organic pollutants such as formaldehyde and benzene, and eliminate the physical absorption saturation and secondary pollution.

Test results by Shanghai Municipal Centre of Disease Control and Prevention show that it kills 99.9% of indoor air germs. Other official tests show that 90% of VOCs (volatile organic compounds) are killed and the dust removal rate can exceed 95%, with a purification efficiency much higher than the national standard.

This multi-functional air cleaner has an LCD panel that can automatically test indoor air quality and show different colours to depict real-time air quality: red means the indoor air quality is bad, blue means it is moderately good and green means it is good.

The intelligent control system of the cleaner can adjust air velocity to match different air qualities and realise energy saving, while keeping the indoor air clean. Its function of releasing negative oxygen ions, at a rate of more than 1m ions per second, can help to purify the air from tiny dust particles.

### China's nanometer green plate technology put into use for printing

According to a report by Beijing Daily on 3 August 2010, nano green plate technology, independently developed by China, was used for magazine printing in August 2010. This was the first time that nano green plate technology was used in the printing of formal national publications after five years' R&D.

Traditional plate-making techniques are based on the principle of photographic imaging. This requires the use of different photographic materials, which can generate substantial wastewater pollution due to the use of many chemicals and reagents during the developing, fixing and washing processes. The nano material eliminates traditional photographic imaging in favour of the digitisation of plate making. Nano green plate technology no longer uses photographic materials. It simplifies the printing process and lowers printing costs, and at the same time it fundamentally changes the printing industry's reliance upon photographic materials. This green and energy-saving technology has replaced laser typesetting and computer-to-plate technology.

Several core patents have been acquired for this technology, along with independent IPRs. Officials from the Chinese Academy of Sciences say that it is a revolutionary technology for the printing industry, representing a shift from "darkness" to "light". (Dong Changqing)

**China successfully develops nano-grating sample plate**

According to a Xinhua News Agency report on 19 November 2010, a research team jointly set up by the National Institute of Metrology, Tongji University and the National University of Defence Technology successfully developed a 1-D Cr Atom Deposition Nano grating Sample Plate for the first time domestically after three years of work.

This plate is one of five nano standard sample plates stipulated by the International Bureau of Weights and Measures. Its successful development fills a gap in China of nano metrology reference material, with technical indicators reaching an internationally advanced level. This has made China one of only a few countries to acquire atom-grating deposition technology. It will resolve the traceability and calibration problem of China's nano metrology instruments and free China from reliance on foreign countries in nano-metrology.

**Chinese scientists develop nano "bionic bone"**

According to a Xinhua News Agency report on 24 March 2011, Professor Tang Ruikang of the chemistry department of Zhejiang University led a research group and developed a new material. Its strength and tenacity are similar to natural bone, realising the biomimetic preparation of nano-scale bone-like microstructure. The results were published in the top materials science journal "Advanced Materials".

Bone is both hard and elastic at the same time. Under an electron microscope, bone reveals exquisite structured layers of HAP crystal embedded in bundles of organic collagen stroma. This multi-layer composite structure gives bone hardness and rigidity. HAP crystal in bone is the thinnest crystal known so far in biological materials, being only 1-2nm thick, which cannot be acquired through conventional experiments. Tang Ruikang said that to produce bionic bone, the biggest challenge is ascertaining whether the organic inorganic composite structure can be simulated to such a precise degree.

Research group members added calcium ion and phosphates to the protein solution modified by different surfactants and then formed white turbid liquid. Under an electron microscope, this liquid shows nano crystal characteristics, but is different from traditional crystals. It is orderly synthetised by around 30 layers of protein and HAP. The organic layer is like the steel frame of a house, while HAP is like bricks and tiles.

Tang Ruikang said that the experiment results showed that each calcium phosphate layer constitutes a complete HAP crystal flake with a thickness of only 2nm, while the thickness of the organic layer is only 1nm. They closely connect neighbouring inorganic layers and keep a certain orientation. Experimenters then used nano probe technology to acquire the mechanical property of nano materials. Under the impact of external forces, the bionic bone could curve to a certain degree while keeping the crystal structure's completeness. When external forces disappear it will bend back to the original shape, without any damage.

According to measurements and calculations, this new material with HAP as the main ingredient has a much higher elasticity than traditional HAP, even exceeding the elasticity of common biological bone. We may call it a kind of elastic crystal, said Tang Ruikang. Currently, the laboratory can also adjust the shape and size of the crystal through changing the ingredients of the bionic bone.

Many domestic and foreign institutions are doing research into bionic bone materials..Once it is put into application, it might help cure bone fractures.

**On 24 March 2011, research group members display the recently developed "bionic bone"**

Wang Dingchang, Xinhua News Agency

## China develops new nano material to safeguard power grid security

According to a Xinhua News Agency report on 2 May 2012, the new power anti-pollution flashover nano composite, room temperature vulcanising silicone rubber material, which was jointly developed by the National Centre for Nanoscience and Technology, the Institute of Process Engineering, the Chinese Academy of Sciences and the China Electric Power Research Institute, provides a strong new weapon for resolving the electric transmission and transformation equipment's pollution flashover.

Pollution flashover occurs during bad weather conditions such as rain, snow, or fog, when pollution such as dust comes into contact with the porcelain insulator of transformer substations or high voltage lines, with ground insulation. This leads to surface creepage and flashover of the insulators, and then causes a tripping operation. As environmental pollution worsens, power systems in countries around the world have frequent accidents brought about by pollution flashover.

Tackling this hidden danger, the National Centre for Nanoscience and Technology, the Institute of Process Engineering, the Chinese Academy of Sciences and the China Electric Power Research Institute developed the new power anti pollution flashover nano composite room temperature vulcanising silicone rubber material between 2006 and 2009. They raised the design thinking of the nano composite, giving full play to the nano integration technology advantage and enhancing the performance of the coating material's hydrophobicity, tracking resistance and ageing resistance.

The power anti pollution flashover nano material has passed several strict inspections and technical evaluations. Scientific research institutions are cooperating to conduct programmes such as nano electric material and application technology research and demonstration and nano insulation composite material inspection technology research. They are striving to carry out the core technology R&D of key insulation materials when nano composite materials are in ultra-high voltage transmission, so as to enhance the insulation grade and service life under a strong electric field to realise the industrialisation demonstration and large-scale application.

## Chinese scientists discover new features of iron-based superconducting materials

Yuan Huiqiu and other scientists at Zhejiang University discovered that 2-D layer iron-based superconducting materials show the superconducting features of

3-D isotropy. The UK's *Nature* journal published this important research result on 29 January 2009 and wrote an introduction in its News & Views column.

This discovery was the first to find the 3-D superconducting feature of 2-D layer superconducting materials, showing that this new iron-based superconductor had different features to the previously researched high-temperature superconductor of copper oxides. *Nature's* experts gave a high opinion of this thesis. They believed that it was an important discovery on superconducting research and has significance for studying the formation mechanism of an iron-based high temperature superconductor.

**China develops second-generation high temperature super conductive belt materials**

According to a Xinhua News Agency report on 23 January 2011, a research team led by Professor Li Yijie in the physics department of Shanghai Jiaotong University adopted a unique technological route and took three years to successfully develop a set of 100-metre-level second-generation high temperature superconducting belt materials with independent Chinese IPR.

These belt materials are like a layer of film, with a metal substrate 1cm wide and 80 micrometres thick. The thickness of the superconductive layer of rare earth oxides used for transmitting super currents is less than 1 micrometer. The current carrying capability of superconductive belt materials is the same as hundreds of traditional copper interconnects with the same cross sectional area.

Across the world, there are only a small number of institutions capable of developing second-generation high temperature superconducting belt materials. Currently, the only domestic scientific research unit capable of producing more than 100 metres of second-generation high temperature superconducting belt materials continuously is Shanghai Jiaotong University. Internationally, since 2004, R&D institutions in the US, Japan and Germany have successfully developed these belt materials that are more than 100 metres long and capable of transmitting super currents above 100A.

# Chapter 5

# Transportation technology

By the end of 2011, China's total expressway length reached 84,900km, ranking second in the world, while its high-speed railway length ranked first in the world.

China's high-speed railway system has reached world-class level. Through construction and operation of the Beijing-Tianjin, Wuhan-Guangzhou, Zhengzhou-Xi'an, Shanghai-Nanjing, Shanghai-Hangzhou and Beijing-Shanghai high speed lines, China has comprehensively grasped the core technologies in areas such as project construction, high-speed trains, train control, station construction, system integration and operation management, and formed complete sets of high-speed technology systems with independent IPR. China has the most complete high-speed railway system technologies, the strongest integration ability, the longest operational mileage, the highest operation speed and the largest amount of track under construction in the world. Its locomotive technology leads the world. Based on grasping the 200-250 kph CRH core technology, China successfully set up the 350 kph CRH technological platform, and developed the 380 kph new-generation, high-speed train. It has systematically grasped the high-power electric and diesel locomotive core technology, and developed the six-axis 7,200kW and 9,600kW high-powered electric locomotives, all with independently owned IPR.

China has also paid close attention to improving its self innovation ability in transportation equipment, such as electric vehicles, traditional vehicles, urban rail transit and ships. It has learned how to digest, absorb and re-innovate introduced technologies, grasped core technologies, and realised the industrialisation of self-owned branded products. It has also developed intelligent technology of comprehensive transportation, enhanced network capacity and transit efficiency, provided a convenient personalised traffic service. Paying attention to energy saving, safety and environmental protection, the country is able to provide complete technical support for the healthy development of sustainable transportation.

## 3D transportation network

### *Railways*

By the end of 2010, China's total length of railway lines in service reached 91,000km, ranking second in the world. Its railway technology has reached world class level. The Qinghai-Tibet railway, which was fully opened to traffic on 1 July 2006, is the world's longest plateau railway with the highest altitude and with the fastest average speed in a permafrost region, thereby promoting the development of permafrost engineering technology. The development of heavy haulage transport is meeting an increase of 200m tons of railway freight each year. On 26 December 2010, the Datong-Qinhuangdao railway broke the world record in terms of heavy haulage, with more than 400m tons of freight carried during the year. China's high-speed railway length ranks first in the world. On 3 December 2010, a new generation of domestic high speed trains, the CRH 380A, created a world-record speed of 486.1 kph on a section of the Beijing-Shanghai line.

#### Qinghai-Tibet railway

The Qinghai-Tibet railway is the highest line in the world. In February 2011, the whole line was successfully made high speed, maintenance-free and unmanned. An operator only needs to click on a computer to initiate safe and fast dispatch of railway transportation.

**The T164 train from Shanghai to Lhasa passes the foot of Nyenchen Tonglha Mountain. On 1 July 2006, the entire Qinghai-Tibet railway was opened to traffic**

Xinhua News Agency

#### Beijing-Tianjin intercity high-speed railway

The Beijing-Tianjin intercity high-speed railway started construction on 4 July 2005 and was put into operation on 1 August 2008. It is 120km long and its highest operational speed is 350kph. It takes about 30 minutes to travel from Beijing to Tianjin, with a minimum running interval of three minutes.

This was China's first IPR-owned high-speed railway and it had the highest operational speed in the world. It connects the two megacities in just 30 minutes, thereby opening the gate for China's railway sector to step into the high-speed era.

Xinhua News Agency

**A test train on the Beijing-Tianjin intercity railway, China's first railway with a speed of 350kph, passes Yongdingmen Bridge in Beijing**

## Beijing-Shanghai high-speed railway

The Beijing-Shanghai high-speed railway is the main north-south line in China's four vertical and four horizontal high-speed railway passenger transport network. It connects the two big economic zones of the Bohai Sea Rim and the Yangtze River Delta. The line begins at Beijing South station, and ends at Shanghai Hongqiao station, by way of seven provinces and municipalities: Beijing, Tianjin, Hebei, Shandong, Anhui, Jiangsu and Shanghai. There are 24 stops on the way, with the minimum travelling time of four hours and 48 minutes. The overall length of the Beijing-Shanghai high-speed railway is 1,318km. Its construction was started on 18 April 2008 and was completed on 15 November 2010. On 30 June 2011, it opened to the public.

Fan Jun, Xinhua News Agency

**On 30 June 2011, a CRH380 high-speed train pulls out of Shanghai Hongqiao railway station. On that day, the Beijing-Shanghai high-speed railway was officially opened to traffic**

## Tunnels and underground projects

By May 2010, China had become the world's fastest developing market for tunnels and underground project construction. Its tunnel construction technology had reached a world advanced level. By 2020, China plans to build 5,000 tunnels, with a total length of more than 9,000km, far greater than the plans for any other country.

### Xiang'an undersea tunnel

On 26 April 2010, mainland China's first undersea tunnel, from Xiamen to Xiang'an, was completed and opened to traffic.

The 8.695km-long Xiang'an Undersea Tunnel is a key project of the national "863" plan. It saves 82 minutes when travelling from Xiamen Island to the mainland. It was designed and constructed independently by China. The tunnel has a subsea zone segment of 6.05km and its deepest point under the water is 70 metres. It took only four years and eight months from the start of the project to completion. Since it opened to traffic, Xiamen Island has formed an all-weather 3D traffic network from sea surface to the sea floor.

### Shiziyang tunnel on Guangzhou-Shenzhen-Hong Kong high-speed railway

On 12 March 2011, the Shiziyang tunnel, which forms part of the Guangzhou-Shenzhen-Hong Kong high-speed railway, was completed. When a train passes through the tunnel, it travels at a designed speed of 350kph, faster than for any other underwater railway tunnel in the world. The 10.8km Shiziyang tunnel is located in the Nansha district of Guangzhou city. It passes under the three water channels of Xiaohuli, Shazaili and Shiziyang.

### Haitangshan tunnel

On 28 March 2011, the longest highway tunnel in northeast China, the Haitangshan tunnel, was completed, laying a solid foundation for the Fuxin-Panjin expressway. The overall length of the tunnel is 7,034 metres, being 3,525 metres long on the left line and 3,509 metres long on the right line. There are four vehicle lanes, eight emergency parking bays and five pedestrian paths. It was one of the most challenging and important construction projects of the Fuxin-Panjin expressway.

### Jiaozhou Bay undersea tunnel in Qingdao

The six-lane Jiaozhou Bay tunnel was officially launched in August 2007 and was completed in April 2010, with a total investment of Rmb7.062bn. The design speed is 80kph, reducing the commuting time between Qingdao and Huangdao to only around 10 minutes. On 30 June 2011, the tunnel was completed and opened to traffic.

### Subway tunnel crossing the Yangtze River

Xinhua News Agency

**On 6 August 2011, after 21-months of hard work by China Railway Tunnel Group, the right line project of Wuhan Subway Line 2 was completed, becoming China's first subway tunnel under the Yangtze River**

In March 2011, Subway Line 2 in Wuhan crossed the Yangtze River. It was China's first subway tunnel crossing the Yangtze River since the Wuhan Yangtze River Tunnel. This subway tunnel is located between Wuhan Yangtze River Bridge and Wuhan Yangtze River Tunnel. It is a cross-river passage of Subway line 2 connecting the Wuchang and Hankou central zones. Its overall length is about 3,100 metres, with an excavating depth of 46 metres. It's a double rail, double-hole tunnel. On 20 September 2011, the left rail project of Wuhan's first subway tunnel, starting from Wuchang on the south bank of the Yangtze River, was successfully dug to Jianghan Road of Hankou district on the north bank. This was another successful crossing of the Yangtze after the right rail project was completed in August 2011.

### Xiangjiang River overpass tunnel

On 23 September 2011, the main construction of the first tunnel across the Xiangjiang River was completed. This tunnel has an overall length of about 8.5km, with a designed speed of 50kph. Four ramps were added along the two banks of the Xiangjiang River. The tunnel crosses the Xiangjiang River eight times and the fault fracture zone 6 times.

## *Highways*

Over the past 10 years, China has accelerated the construction of expressways. The trunk lines of five vertical and seven horizontal national highways were all completed 13 years ahead of schedule. Eight interprovincial highways that formed part of the Western Development programme were completed. A highway traffic network was initially formed, covering both urban and rural areas, and it has proved to be convenient and highly efficient. From 2002 to 2011, China's total highway length grew from 1.7652m km to 4.1064m km, of which the length of expressway roads surged from 25,100km to 84,900km, rising to second place in the world.

**Bohai Sea Rim Beach expressway completed**

According to a Xinhua News Agency report on 30 November 2011, the opening to traffic of Tianjin Beach Road marked the completion of China's Bohai Sea Rim Beach expressway.

The beach expressway has a total length of 302km. It starts from Beidaihe Service Zone and connects to Tianjin Beach Road to the west.

The total length of the Tianjin Beach Road is 91km, passing through the whole of Binhai New Area. It starts from Jianhe in the Tangshan section of the Beach Expressway in the north, and then goes south to the Cangzhou section at Qikou via several large economic zones of Binhai New Area, including Nangang Industrial Zone, Binhai Tourism Zone and the Central Fishing Port Zone. It will become the main traffic artery of Bohai Sea Rim Economic Zone, significantly reducing traffic pressure in the central urban district.

This four-lane expressway has a subgrade that is flush with the surrounding ground, and it has a 28-metre-wide roadbed and a design speed of 120kph. It is totally enclosed, all overpassed, with a total length of 160km. After opening to traffic, it should take only 90 minutes to travel from Tianjin Binhai New Area to Beidaihe.

The beach expressway connects Qinhuangdao port, the Caofeidian Harbour district of Jingtang port, Tianjin port, Huanghua port and Jingtang port, as well as several tourist attractions. It promotes connections between these ports, and is of great significance for constructing the new traffic pattern of "eastward exit and westward connection". It has played an important role in the economic development of the Beijing-Tianjin-Hebei region.

**First six-lane expressway for west-east coal transportation completed**

On 16 December 2011, the first six-lane expressway for west-east coal transportation was completed and opened to traffic in the northern province of

Shaanxi. From Shenmu county to the Shaanxi-Shanxi border in Fugu county, the road has shortened the journey by nearly two hours to around 40 minutes.

The Shenmu-Fugu expressway is a key section of the second high-speed road, the Yulin Shangnan line of the Shaanxi Provincial Expressway Network connecting the three economic zones of northern Shaanxi, Guanzhong and southern Shaanxi. It is also the only high-speed road for the eastward transportation of coal in Yulin Shenfu coal field, which is a national energy base. The main part of the Shenmu-Fugu expressway is 56.9km long and has a designed speed is 80kph. It has six lanes and involved a total investment of Rmb7.45bn.

The Shenfu field is a major source of high quality coal. For a long time it was mostly transported by road. In recent years, with the rapid development of the domestic economy and a surge in road traffic, highway congestion has become the norm. The completion of the Shenmu-Fugu expressway provides a high-speed outward transport channel for western coal transportation.

**The Yamane-Xichang expressway**

Jiang Hongjing Xinhua News Agency

**A corner view of Yaxi expressway's Ganhaizite Bridge**

On 28 April 2012, the opening ceremony of the Ya'an-Xichang expressway was held near the Ganhaizite Bridge. This marked the completion of the hardest and most magnificent section of the Beijing-Kunming expressway.

The 240km-long Ya'an-Xichang expressway is a key component of the Beijing-Kunming expressway (G5) and the Gansu-Lanzhou-Yunnan-Mohan highway. The people of Sichuan province in southwest China have been waiting for this project for many years. It was launched in 2007, and involved an investment of about Rmb20.6bn.

The altitude of the Ya'an-Xichang expressway rises from the edge of the Sichuan Basin to the Hengduan Mountain Region. It crosses a remote mountain canyon of southwestern China where geological disasters take place frequently. The construction of the expressway involved overcoming a challenging natural environment and high degree of engineering difficulty. The Labajinte Bridge is the highest bridge pier of its type in the world; and the Ganhaizite Bridge was the first in China to use the lightweight steel tube truss continuous beam.

### Suzhou-Nantong bridge opens to traffic

On 17 April 2008, the Suzhou-Nantong bridge, a cable stayed bridge with the longest span in the world, was opened to traffic. Independently designed and constructed by China, the bridge connects the Jiangsu cities of Nantong and Suzhou. Its total length is 32.4km, and its main span is 1,088 metres long. The bridge tower is 300 metres high, creating four world records for a cable-stayed bridge: the deepest pile foundation, the highest cable bent tower, the widest span and the longest stay cable.

### Qingdao Jiaozhou Bay bridge opens to traffic

On 30 June 2011, the Qingdao Jiaozhou Bay bridge was officially opened to traffic. It was the first bridge cluster project over the frozen waters of north China. It is the starting point of the Qingdao-Lanzhou expressway and has a total length of 36.48km, making it the world's longest cross sea bridge. The bridge starts in a high-tech park in eastern Qingdao, crosses Jiaozhou Bay, and connects to the south line of the Jinan-Qingdao expressway at Huangdao, connecting Qingdao, Huangdao and Hongdao in the shape of the Chinese character "品". The bridge includes a six-lane expressway and an eight-lane city expressway, with a design speed of 80 kph, a width of 35 metres, and a designed lifespan of 100 years.

**An aerial photo of Qingdao Jiaozhou Bay Bridge taken on 21 June 2011**

## New-generation high-speed train

On 26 February 2008, the Ministry of Science and Technology and the Ministry of Railways jointly agreed to cooperate on the joint innovation of a Chinese high-speed train. The overall objective was to develop a new-generation high-speed train that can travel at 350kph or above. So far, breakthroughs have been made in the following 10 key technologies: aerodynamics, transmission and brakes, operational control, running gear structure, vibration and noise reduction, traction power supply, lightweight vehicle body, comfort, security intelligence and transportation organisation.

### New-generation CRH-380A rolls off Changchun assembly line

On 27 May 2010, the first CRH-380A train went off the assembly line at the high-speed train manufacturing base of CNR Changchun Railway Vehicles. It is a new-generation high-speed train with an independent IPR. The CRH380A is designed to operate at a cruise speed of 350kph and a maximum of 380kph in commercial service, making it the fastest train in the world.

The CRH380A has a streamlined head and a highly pressurised tight body. Its advanced noise-control technology and high-performance traction system make the train operate in a steady and quiet manner, thereby saving energy and making it more environmentally friendly.

### Higher-speed test train completed

In November 2011, research and development into high-speed CRH wheels, one of the national "863" projects, was verified and accepted by experts. The expert assessment team believed that the project designed the wheel tread profile appropriate for Chinese railways. The wheel's comprehensive performance at least matched the level of imported wheels from Europe. In December, a higher-speed test train was completed at CSR Sifang. This signalled that China had achieved significant results in the study of high-speed trains, and should give China a greater global profile in high-speed railways.

This test train is based on the innovative results of CRH 380A. Its priority is safe and reliable operation at higher speed. Focusing on enhancing critical velocity, tractive capability and reducing resistance, the test train incorporates innovative systems integration, head type, vehicle body and bogie, propulsion and braking systems. It has a grey body, black stripes in the middle, and "CRH" logo. The head looks like "a drawn sword", and the tail is shaped like a rocket.

Technical staff stated that the trial speed of the test train will be higher than the real maximum operation speed of the current Beijing-Shanghai high-speed trains. It will undergo 11 kinds of tests. The train itself contains more than 1,100 sensors, of which 545 are used in the tests. Sci-tech researchers have studied factors affecting safety, energy saving and passenger comfort while the train operates at high speed.

This test train has six compartments, all of which are motor vehicles. Its intelligent level on security monitoring and fault diagnosis has improved. New technologies such as Ethernet Ring control, air resistance brake and new materials such as carbon fibre, magnesium alloy and the new nano acoustic insulation have been successfully applied.

## Maglev train

A maglev train is levitated on top of a guideway using magnets to create both lift and propulsion.

The maglev train is the only large volume, energy-saving and environmentally friendly type of ground passenger transportation that can operate safely and economically at a speed of 400-500kph. It is much more energy efficient than other vehicles and aircraft travelling at a similar speed.

### China's first domestic high-speed maglev sample train delivered in Chengdu

On 8 April 2010, China's first domestic high-speed maglev sample train, manufactured by AVIC Chengdu Aircraft Industrial (Group), was delivered in Chengdu. This showed that AVIC Chengdu had the ability to design, manufacture and integrate a maglev train.

### Chinese-owned maglev line starts construction

On 28 February 2011, construction started on a medium- and low-speed maglev transport exemplary line (S1). China will own its first maglev line with independent IPR; it will be the world's second medium- and low-speed maglev line in operation.

### Medium- and low-speed maglev line to settle in Zhangjiajie

On 1 April 2011, Hunan province's Zhangjiajie government, CSR Corporation, China Railway Group and Hunan Provincial Railway Investment signed an agreement to build a medium- and low-speed maglev commercial line between the urban area of Zhangjiajie and a World Natural Heritage Site in Wulingyuan by the end of 2012. Its maximum operation speed would be about 120kph and it can transport up to 402 people.

**Chinese maglev train rolls off the assembly line**

On 20 January 2012, a medium- and low-speed maglev train developed by CSR Zhuzhou Electric Locomotive was completed in Zhuzhou, Hunan province. The maglev train can reach a maximum speed of 100kph and can carry about 600 passengers. Its characteristics, including lack of track friction, zero dust emission and low noise, will meet the high environmental standards.

Currently, only a very few countries have managed this technology, and there is only a small number of commercial trains in existence.

Li Ga, Xinhua News Agency

On 20 January 2012, a medium- and low-speed maglev train rolls off the assembly line of CSR Zhuzhou Electric Locomotive

## New energy vehicles

New energy vehicles have new technology and structures that are powered by non-conventional fuel (or, using conventional fuel, adopt a new type of vehicle-mounted power unit), and synthesises advanced power control and drive vehicle technologies.

The power of a new energy vehicle includes all energy sources except petrol and diesel. It mainly concerns pure electric vehicles, hybrid electric vehicles and fuel cell electric vehicles. The driving forces powering these three types are charging, the internal combustion engine and transformation of chemical energy. A new energy vehicle consists of the car body, electric drive and control, energy storage battery, and energy management.

## Top three types of new energy vehicles: pure electric, hybrid and fuel cell

**Comparison of features**

| Vehicle type | Power resource | Features |
|---|---|---|
| Pure electric vehicle | Storage battery power | 1. Electric motor takes place of internal combustion engine<br>2. Zero emissions |
| Hybrid vehicle | Internal combustion engine plus battery power | 1. Electric motor and internal combustion engine jointly provide power<br>2. Engine keeps the best comprehensive performance, lowers oil consumption and emission |
| Fuel cell vehicle | Chemical energy | 1. Electrochemical reaction takes place of internal combustion engine<br>2. Emission is water<br>3. Needs to bring special fuel |

The pure electric vehicle is also called a battery electric vehicle, or a secondary battery electric vehicle. The main distinction between pure electric vehicles and traditional vehicles is that the electric motor takes the place of the internal combustion engine, providing power by a rechargeable battery.

The hybrid vehicle is equipped with two power sources, heat power (generated by a traditional petrol or diesel engine) and electric power (battery and electric motor). By using an electric motor, a hybrid vehicle's dynamic system can be flexibly controlled according to the real operation situation, while the engine keeps working in an optimal area of comprehensive performance, so as to lower oil consumption and emissions.

The power of the fuel cell electric vehicle comes from the transformation from chemical energy to electric energy. This kind of vehicle must carry special fuel, and use the electric energy generated by hydrogen and oxygen in the electrochemical reaction of the fuel cell in the presence of a catalyst. The earliest fuel cell electric vehicles used hydrogen fuel directly, with the storage of hydrogen in the form of liquefied or compressed hydrogen, or metal hydride. The hydrogen vehicle only emits pure water and therefore has zero pollution, zero emissions and abundant reserves.

## Government subsidy for buying new energy vehicles

Four ministries and commissions, including the Ministry of Finance and the Ministry of Science and Technology, have pledged that the government will subsidise the purchase of new energy vehicles in 13 pilot cities. The goal is to enlarge the market and encourage enterprises to produce, research, and develop energy-saving and new energy vehicles.

As part of this pilot scheme, funds will be used to encourage and promote the use of energy-saving and new energy vehicles in public service areas such as public buses, official business, environmental sanitation and the postal service. The central finance department issues a one-off subsidy to units in the public service area that buy and use energy-saving and new energy vehicles. An allowance is also given for the construction and maintenance of relevant facilities.

Demonstration and promotion units must optimise purchases through public bidding, and stipulate the vehicle type, quantity, price and after-sales service required.

# Chapter 6

# Energy and environmental protection

In 2011, China's installed hydropower capacity exceeded 230m kW, ranking first in the world. Fifteen nuclear power units have been completed and put into operation, with an installed capacity of 12.538m kW. China's nuclear power capacity under construction accounts for more than 40% of the world total; the installed capacity of windpower totals 47m kW, ranking first place in the world; the utilisation of solar water heaters ranks first in the world, with a collector area of 180m sq metres; the installed photovoltaic (PV) capacity reached 3m kW, a growth of at least three times more than the previous year. The production of PV cells accounts for half of the global market; the installed capacity of biomass power generation has reached 4.5m kW, raising the proportion of non-fossil fuels in primary energy consumption to 8.1%.

China now produces all washing appliances for coal preparation plants with a capacity of 4m tons a year. Technologies, such as dense medium coal separation, have been widely applied. China has started to develop direct coal liquefaction technology with an independent IPR. A direct liquefaction production line that can generate 1m tons a year has been put to the test. Large volume and high-parameter thermal power generating units have been completed and gone into operation. The number of 600°C ultra-super critical units now tops the world, with the efficiency of generating units being more than 45%. A 1,000MW direct air cooling unit with an independent IPR has been put into operation. A subcritical 300MW circulating fluidised bed (CFB) boiler is in mass production. A supercritical 600MW CFB boiler is under development. The key technologies of 100kW and 1MW gas turbines for a distributed heat electricity cool cogeneration system have made progress. The construction of an exemplary project has begun for a 250MW integrated gasification combined cycle (IGCC) unit with the gasification technology of an independent IPR. An AC-testing project of 1,000kV high voltage transmission and an ±800kV DC have been successfully put into operation. Research on an intermittent power grid and an energy storage technology has achieved some encouraging results.

## The development of a large oil and gas field and coalbed methane

During the 11th five-year plan, a special oil and gas development project investigated exploration, development engineering, technology and coal bed methane, forming 21 new theories such as litho-stratigraphic reservoirs. Twenty-four major core technologies and 162 proprietary technologies have been identified, such as enhanced oil recovery from high aqueous oilfields, a high-temperature and high-salinity reservoir and the high-efficient development of offshore heavy oil. A total of 115 sets were developed, such as the integration of seismic processing and interpretation; 934 patents were applied for, of which 637 were patents for invention. These achievements have enhanced the independent innovation ability of key technologies for exploring and developing China's oil, gas and coal bed methane, and realised a great leap forward in the development of natural gas. China has achieved major progress in the following areas.

### Petroleum geology and exploration technology

Systematic breakthroughs have been achieved in formation conditions, distribution characteristics, reservoir forming regularity and exploration technology of large oil and gas fields, which guide the rapid development of eight major oil and gas areas including Erdos and Sichuan Chuanzhong.

### Enhanced oil recovery technology in high water cut oil fields

New technology on enhanced oil recovery of high water cut oil fields has ensured that the Daqing Oilfield generates a steady yearly production of 40m tonnes crude oil for five consecutive years.

### Major equipment development

Jin Liangkuai, Xinhua News Agency

**HYSY-981 is the first sixth-generation deep-water drilling platform that was independently designed and constructed by China. Its maximum operating water depth is 3,000 metres and its maximum drilling depth is 10,000 metres. The weight of the platform is more than 30,000 tons; from the bottom of the vessel to the top of the platform, the span is 137 metres, equivalent to a 45-storey building**

China has successfully developed a 10,000-channel large seismometer prototype. Its data transmission ability is two to five times higher than international products of a similar type. Its performance indicators and data quality of field acquisition have reached international level.

China has successfully developed a 3,000-metre deepwater semi-submersible platform. It is the first top-level deepwater semi submersible platform to have been completed in China. The breakthrough means that China has realised a significant increase in depth from 500 metres depth to 3,000 metres.

China has successfully developed about 120 sets of the Epilog system. They have been put into production in domestic and foreign oilfields, undertaking 50,000 well logging operations and greatly improving the accuracy of recognising complicated oil and gas reservoirs logging, with a 30% improvement in logging efficiency, making technical indicators and product quality reach an internationally advanced level.

**Low-cost production technology of coalbed methane**

China's coal bed methane entered a new stage of industrialisation with the establishment of an enriched and accumulated model of the Qinshui coal bed methane field and set up the first digital large-scale coal bed methane field in the south of Qinshui Basin.

## Water pollution control and treatment

### *Interpretation of key terms*

To tackle increasingly severe water pollution, China launched a major water body pollution control and treatment project in 2007 to implement its strategy of pollution source control and emission reduction. During the $11^{th}$ five-year plan, this special project concentrated on about 600 pollution control technologies in five key industries: chemical engineering, light industry, metallurgy, textile dyeing and printing, and pharmacy. During the $12^{th}$ five-year plan, China has changed many of its ideas on pollution prevention and control: the transformation of national water pollution control and treatment from prevention and control of pollution to prevention and control of pollution and ecological restoration.

The special project takes four rivers, the Liaohe, Huaihe, Haihe and Songhuajiang, and three lakes, Taihu, Chaohu and Dianchi, along with the Three Gorges reservoir as key implementation areas. The implementation is divided into three stages over 15 years (2006-20). During the implementation, six themes are designed: lake eutrophication control, water treatment, comprehensive treatment of city water, safeguarding drinking water, monitoring and early warning and environmental strategy and policy research.

## *Special achievements*

### Six hundred breakthrough pollution-control technologies

During the 11th five-year plan, this special project concentrated on some 600 key pollution-control technologies in five key industries: chemical engineering, light industry, metallurgy, textile dyeing and printing, and pharmacy. All of them have received engineering verification in 139 large projects. Through demonstration in key river basins of the Liaohe, Haihe and Songhuajiang rivers, 130m tons of sewage and 11,000 tons of chemical oxygen demand (COD) are reduced a year; key technologies have been developed, such as upgrading and reconstructing urban sewage treatment plants and In-depth denitrification and dephosphorisation; about 500 urban sewage treatment plants have been upgraded; every year, 160,000 tonnes of COD, 54,000 tonnes of ammonia nitrogen, 14,000 tonnes of total phosphorus are reduced, playing an active role in improving the water environment of river basins.

In addition, an engineering technoligal solution for safeguarding drinking water has been initially established from water source to tap. Aimed at prominent water pollution problems in important regions, research has started in technologies designed to improve water quality, which has greatly reduced the burden of water source pollution. They are investigating key technologies such as the removal of microsystems by advanced oxidation, combined disinfection of ozone and biological activated carbon and UV, rebuilding secondary water supply, pipeline network risk control, water quality conservation, and optimum anti-seismic design of the water supply system in earthquake-prone regions. More than 40 key technologies for drinking water safeguards have been developed; and 82 technological breakthroughs have been realised. Twenty-five new materials and equipment on water quality detection and purification have been developed, while 31 projects/R&D bases have been established to ensure water quality and drinking water safety in water supply plants. By implementing key technology in pollution source control and emission reduction, which has been applied in nearly 600 exemplary projects, the deterioration of water quality in key rivers has been curbed, with an improvement of water quality in some river sections and basins.

**This is a photo of a "green floating island" for treating water pollution in the Daguan River, taken in Kunming city, Yunnan province, on 8 October 2009. These green plants in the shape of "floating islands" beautify the environment, and can absorb substances in water, such as nitrogen and phosphorus, so as to restrain blue-green algae outbreaks**

Mei Jianguo, Xinhua News Agency

Environmental monitoring shows that, in terms of COD in the main streams of the Liaohe and Huaihe rivers, Grade V water has been eliminated. In 2010, the average water quality in the comprehensive demonstration zone of Songhuajiang River Basin reached Grade IV. Water quality in the Haihe River has also been improved. The eutrophication level of Taihu Lake changed from medium to light, with the number of Grade V inflowing rivers decreasing from eight to one, the number of inflowing rivers worse than Grade V decreasing from three to two, the number of Grade III inflowing rivers increasing from one to three. The eutrophication level of Chaohu Lake has improved remarkably, restricting the widespread incidence of cyanobacteria bloom. In Dianchi Lake Basin, the Panlong River, along which population density is high and the pollution burden is heavy, black and odorous water has been eradicated. Its main water quality indicators are better than Grade V for river channels. Its water environment has been improved fundamentally, with a clear improvement in the outer part of Dianchi Lake.

**R&D of 50 key technologies and industrialisation equipment urgently needed**

Jilin Chemical Fibre Group produces 136,000 tons of acrylic fibre a year, accounting for 16% of China's total production capacity, making it the second biggest acrylic fibre manufacturer in the country. The technology of physico-chemical-bio treatment of synthetic fibre manufacturing wastewater that has been used in the enterprise reduces the cost of wastewater treatment by more than 30%. Each year, 310 tons of toxic dimethylacetamide (DMAC) are reduced from the emission, which has essentially resolved the twin difficulty of removing degradation-resistant toxic organic substance and denitrification in the treatment of synthetic fibre wastewater. Aimed at resolving the problem of a large quantity of wastewater and treatment difficulty, Changchun Dacheng

Group developed the manufacturing technique of electro dialysis desalination, a fermented culture whose acid yield rate and conversion rate reaches an internationally advanced level and fermentation engineering technology. If it is promoted across the whole industry, 3m tons of water can be saved every year, while cutting down COD.

Because China was not advanced in the design and production of water environment monitoring, sludge treatment and water treatment equipment, the country developed 50 key technologies and industrialisation equipment that were urgently needed, assisted a group of green enterprises to go public and cultivated an output value of around Rmb4bn for the environmental protection industry.

In a comprehensive demonstration zone of an industrial park in upstream Meiliang Lake within the Taihu Lake Basin, the implementation of water body pollution control and treatment programmes helped government departments to shut down or move 454 enterprises. There are 116 enterprises implementing cleaner production, cutting pollutants by about 10% and saving 1.3m tons of water a year. Based on the key production technology and demonstration project of recycled water, the zone has a yearly production capacity of 24m tons of recycled water and has built up a circular economy system and optimised its industrial and economic structure. (by Li He)

(*Science and Technology Daily*, Page 13, 8 March 2012)

## Clean coal technology

### High-efficient clean power generation technology

- For China's first IGCC demonstration project, under construction in Tianjin Binhai New Area, the main equipment has entered the site for Huaneng Group's 250MW IGCC demonstration power station for its green coal power generation engineering. This marks substantial progress in China's green coal power technology plan as it strives to become a world leader in this area.

IGCC is an advanced power system combining the gasification technology of purifying coal and the efficient combined cycle. It is an advanced coal power generation technology recognised both at home and abroad. With superb environmental performance, the emission load of pollutants is only one-tenth that of conventional coal-fired power plants. Its desulphurisation efficiency can reach 99%, and its nitric oxide emission rate is only 15-20% that of conventional power stations. In addition, zero emissions are relatively easier to realise through this technology.

According to incomplete statistics, there are five large-scale IGCC power stations that have been put into operation using coal as their main raw material. Their total installed capacity is about 1.3m kW. There are more than 40 IGCC power stations under construction or in the planning, with a total installed capacity of about 20m kW. They are mainly located in the US, Europe, Japan and other developed countries. Meanwhile, 17 power stations in the world are planning to adopt "green coal power" technologies to realise near-zero emissions for coal power generation based on IGCC, of which nine are in the US and three in the UK.

- By the end of 2010, the number of supercritical units that have been in operation in China accounted for half of the world's total, with a continuous decrease of coal consumption by thermal power generating units. Independent design and production have been realised for the main components of 1,000 MW units.
- The world's first 600 MW supercritical large CFB boiler has begun construction.

**Coal conversion technology**

- On 18 March 2010, a multi-nozzle opposed coal slurry gasifier processing 1,960 tons of coal a day in a single furnace entered a long period of stable operation.

Huitengliang wind farm in Xilingol league, Inner Mongolia. The photo was taken on 21 December 2011

Ren Junchuan, Xinhua News Agency

- On 8 August 2010, goods were commissioned to supply a 600,000-tons-a-year MTO (methanol to olefin) owned by Shenhua Group, putting China at the forefront of the world's coal-based olefins industry.
- In 2010, a demonstration project of Shenhua megaton direct coal liquefaction entered full operation in Erdos, Inner Mongolia.

## Windpower generation

### *Interpretation of key terms*

Wind power generation involves wind driving rotating windmill blades, then enhancing the rotating speed through a booster engine to make the generator produce electricity. It is a renewable, non polluting source of energy that has attracted strong interest across the world.

On 6 July 2010, the 34 wind turbine generators of the demonstration project of Shanghai East Sea Bridge 100MW offshore wind plant all joined the power grid. This was a milestone in China's offshore wind power development. In September 2010, China's first two 1.5MW high-altitude wind power generators rolled off the assembly line.

In 2011, China consolidated its leading role in global wind power generation by recording an installed wind power capacity of 62,700MW.

### *Sci-tech serves people's lives*

In 2008, projects for the Beijing Olympics included 33 1.5MW wind turbine generators providing clean electricity. Each year, 100m kWh can be produced, which could meet the power needs of 100,000 homes, if the average yearly consumption per Beijing family is 1,000 kWh.

During the 2010 Shanghai World Expo, 34 3MW wind turbine generators of Shanghai East Sea Bridge provided clean electricity for the Expo,

## Solar photovoltaic power generation

### *Interpretation of key terms*

Solar photovoltaic power generation transfers the radiant energy of sunshine into electricity directly by means of the photovoltaic effect of semiconductor

15 February 2012, Ningxia Tianjingshan photovoltaic power station

Li Ran, Xinhua News Agency

materials in solar cells. It can be operated in two ways, either independently or by grid connection. A large photovoltaic power station is well suited in areas of rich solar energy, large deserts, idle land resource and good synchronised conditions. It is emerging as a key part of China's new energy. A storage battery is needed as an energy storing device of the independently operated photovoltaic power generation system, which is mainly for outlying areas and sparsely inhabited areas and therefore involves a high cost of construction.

In the area of photovoltaic power generation application technology, domestic and foreign levels are equally matched. China encourages the R&D and application of design and integration technology for large terrestrial photovoltaic (PV) systems. This provides the driving force for scientific research innovation of the photovoltaic industry.

- On 28 December 2010, China's first demonstration project of a large PV franchise, Dunhuang 20MW PV power generation project was connected to the grid and power generation.
- In 2011, a 1,000MW PV power station was completed and connected to the grid in Qinghai Qaidam Basin, which became the most concentrated area in the world for PV installed capacity.
- On 22 December 2011, the 30MW Ningxia Tianjingshan PV power station near the Tengger Desert was successfully connected to the grid and began power generation. This power station can follow the sun around the clock. Compared with a traditional PV power station, its electricity output can increase by around 25%, with an average annual power generation of 60m kHz and an annual output value of more than Rmb60m.

## *Sci-tech serving the people*

### PV power generation at Shanghai World Expo

The solar PV power generation system at Shanghai World Expo Park was the largest in World Expo history. The system displayed the application of clean energy PV power generation in our daily life. The roof and glass curtain wall of landmark buildings at the Expo Park, such as China Pavilion, Theme Pavilion, Expo Centre and Culture Centre, were installed with a 4.68MW solar power generation system, which can reduce carbon dioxide emissions by 4,000 tons. It was the largest domestic PV cell demonstration zone. In particular, the roof area for PV cells in the Theme Pavilion totalled more than 30,000 sq metres, making it the largest PV roof of a single building in the world.

### Beijing's largest PV roofing power generation project completed

On 1 March 2012, the BOE G8.5 TFT-LCD production line project was completed and went into operation. It was Beijing's biggest "Golden Sun" national demonstration project. It was not only Beijing's largest scale PV roofing power generation project, but also the largest scale PV roofing demonstration project in an industrial factory building among the 13 concentrated application zones of national PV power generation. The project's total installed capacity is 5,000 kW, saving 3,200 tons of coal a year and reducing carbon dioxide emissions of 6,240 tons a year. Its environmental benefit is very significant.

### Benefits for vegetable cultivation and power generation in solar PV greenhouses

Pudong town, located in the city of Jimo, Shandong province, invested more than Rmb20m to build six PV greenhouses. The roofs are installed with thin-film solar cell panels. Inside the greenhouse, an LED nano supplementary lighting system can provide plenty of illumination for vegetables during cloudy conditions, or at night, so as to shorten the vegetables' growth cycle and improve productivity and quality. It is reported that each greenhouse is equivalent to a 200kW power station, which can generate 280,000 kWh and yield a benefit of Rmb3m. Each greenhouse occupies an area of 4,670 sq metres. If the net profit of greenhouse crops is calculated at Rmb25,000 a year, the annual agricultural income could reach Rmb175,000, much higher than can be achieved in traditional greenhouses.

### Solar bus shelter unveiled

On 23 May 2011, the first 14 solar bus shelters that are energy-saving and environmentally friendly were installed and put into use in Lianyungang, Jiangsu

Geng Yuhe, Xinhua News Agency

On 23 May 2011, on Huaguoshan Road in Lianyungang city, Jiangsu province, a public bus stops at a solar-powered bus shelter

province. Each bus shelter comes with a solar monocrystal silicon accumulator plate, enough to provide plenty of electric power for the LED and billboard in the bus shelter.

## Biomass energy

### *Interpretation of key terms*

Biomass encompasses several categories including plants, animals and their excrement, rubbish and organic wastewater. Broadly speaking, biomass is the organics generated from the photosynthesis of plants. Its energy originally came from the sun and therefore, biomass is one kind of solar energy. Bioenergy is a new type of environmentally friendly energy. It is not the same as conventional fossil energy, but nor is it the same as other new forms of energy. It combines the characteristics and advantages of both and is a main renewable energy.

The carrier of biomass is organisms. Therefore, this kind of energy exists in the form of material objects. It is the only type of renewable energy that can be stored and transported. In addition, it is widely distributed, and not limited by weather or natural conditions. As long as there is life, there is biomass. In terms of the utilisation method, biomass has a similar structure to coal and oil and can be processed and used through the same or similar technology. Therefore its utilisation technology has been relatively easy to develop and spread. Moreover, biomass can be converted into electricity, oil, gas or solid fuels, which are directly used by vehicles or conventional thermal equipment. It can be used in almost all aspects of current human industrial production or social life. Therefore, of all the new energies, biomass is quite compatible with modern industrialised technologies and modern life. With a strong substituting capacity, it can take the place of conventional energies without the need to improve on existing industrial technologies. These are the foundations for biomass to play an important role in future. As a rising star in the area of new energy, the types of biomass that can be used as energy include straw, algae, methane, wood and livestock manure,. These energy resources have a wide distribution, a huge development potential, a small environmental impact and are fit for sustainable utilisation.

### *Sci-tech serving the people*

In China, because biomass is currently mainly used in the rural economy, the situation of rural energy demand and consumption in the future will have a great influence on the quantity of biomass development and utilisation.

China has developed several kinds of fixed-bed gasifiers and fluidised bed gasifiers, using stalks, sawdust, rice husks and twigs as raw materials to produce fuels.

The Chinese government and relevant authorities attach great importance to the utilisation of biomass energy and carry out the R&D of the utilisation tech nology of biomass energy, such as a household biogas digester, a firewood-saving stove and a kang (a heatable brick bed used in north China), a firewood forest, a large and medium-sized biogas project, biomass briquetting, gasification and gasification power generation, biomass liquid fuel and gained excellent results.

## Nuclear energy

### *Display of achievement*

#### China develops GW-level full-scale simulator in nuclear power plant

On 28 December 2010, China's first GW-level full-scale simulator in a nuclear power plant was put into use at Fujian Ningde nuclear power station. This marked another important step forward for Chinese independent nuclear power design, manufacture, construction and operation, ending the situation whereby China relied heavily on imports. The full-scale simulator is called a "virtual nuclear power station". According to international practice and Chinese nuclear power station construction standards, each newly built nuclear power station should be equipped with at least one full-scale simulator for carrying out controller training on key nuclear power equipment and licensing examination, to secure the safe operation of the station.

**This is a first stage project of Fujian Ningde nuclear power plant under construction**

## China's first fast reactor connected to grid

China's first experimental fast reactor (FBTR), in which nuclear fission reaction is triggered by fast neutron, was combined to the grid and began to generate electricity at 10am on 21 July 2011. The realisation of this major project of the national "863" plan marked a major breakthrough for fast reactor technology, which is listed as the forefront technology in the national medium- and long-term programme for science and technology development. This also marked an important step forward for China to gain ascendancy in nuclear energy technology and establish an advanced nuclear energy system with sustainable development.

This fast reactor adopts an advanced pool type structure, with a thermal power of 65MW and an experimental power of 20MW. It is one of the few high-power experimental fast reactors with a power generation function. It shares the same main system settings and preferences as large fast reactor power stations. The experimental fast reactor makes full use of inherent safety features and adopts several kinds of passive safety technologies, making its safety level reach the requirements of fourth-generation nuclear energy systems.

The fast breeder reactor is the main type of fourth-generation nuclear energy system. There are two major advantages of the advanced nuclear fuel circulatory system drawn by the fast reactor: one is that it could greatly enhance the utilisation rate of uranium resources, which can be raised from about 1% in a pressurised water reactor that has been widely applied in current nuclear power plants, up to more than 60%; the other is that it can transmutate long-life radioactive waste generated by the pressurised water reactor, minimising the quantity of waste. The development and spread of fast reactor technology have significance in promoting the sustainable development of nuclear power in China, the establishment of an advanced fuel circulatory system and the sustainable development of nuclear energy.

China's fast reactor project is regulated by the Ministry of Science and Technology and the State Administration of Science, Technology and Industry for National Defence. It is organised by the China National Nuclear Corporation and implemented by the China Institute of Atomic Energy (CIAE). For many years, the CIAE has coordinated several hundred units including domestic universities, research institutes and enterprises. After continuous innovative exploration and collaborative research for many years, CIAE completed the research, design, construction and debugging. In May 2009, the first system thermal debugging was started. On 21 July 2010, the first nuclear criticality was realised.

**China's 3rd generation nuclear power becomes safer and more efficient**

The world's first AP 1000 nuclear power unit, which was also China's first nuclear power project, is expected to be connected to the grid and generate power by the end of 2015; the safety analysis report of China's first IPR-owned CAP1400 third-generation nuclear power unit was submitted in March 2012 and construction is expected to begin in 2015; in future, China will also develop 170 GW CAP1700 nuclear power technology.

In recent years, China has continued to set new records in nuclear power construction, break through the domestication bottleneck of key equipment and take firm steps in forming a nuclear power brand with an independent IPR. The self reliance of third-generation nuclear power is the mainstream of global nuclear power development, with good safety and economic prospects, a short construction cycle, long service life and other features.

AP1000 is the universally acknowledged, safest third-generation nuclear power technology. But due the manufacturing challenges involved in the AP1000 equipment, the domestication of key equipment is China's main bottleneck in its self reliance on nuclear power.

After the introduction of the AP1000 technology, and by overcoming various difficulties, China was able to grasp the five major nuclear technologies: nuclear island containment, module, main pipe, heavy forging manufacturing and the one-off pouring of raft foundation concrete for the nuclear island. China also collaborated with parties at home and abroad and finished the development of all the key equipment of the AP1000. The world's first AP1000 nuclear power unit is expected to be connected to the grid and begin to generate power by the end of 2015. Meanwhile, the State Nuclear Power Technology Corporation is looking at how to develop CAP 1400 nuclear technology with a self owned IPR, so as to meet the future needs of nuclear power.

### China supplies high/low pressure heaters to Ningde nuclear station

On the morning of 18 April 2012, six high/low pressure heaters of the No.3 unit for CGNPC Ningde•Dongguo 100 GW nuclear power plant, which are designed and manufactured by Dongfang Boiler Group of Dongfang Electric Corporation, started their journey to Ningde nuclear power plant. This was China's first successful development of GW-level high/low pressure heaters.

The PWR nuclear power plant mainly consists of a nuclear island and conventional island. The high/low pressure heaters are key equipment for the thermal system of the conventional island. They play an important role in improving the unit's cycle efficiency, reducing the corrosion of oxygen and other non-condensed gasses to the main equipment of the nuclear island vapour generator, auxiliary machinery of the thermodynamic system and pipeline valve. It has strict design and manufacturing technology requirements. According to the introduction, the GW-level high/low pressure heater has the highest volume of similar equipment in the world.

### Chinese nuclear power enterprises grasp automatic welding technology for main pipeline

According to a Xinhua News Agency report on 27 April 2012, the automatic welding of the main pipeline at unit No.1 was smoothly completed for the Guangdong Yangjiang nuclear power project. Along the critical path, the real welding has a construction period of 88 days with a 100% qualified rate of welding.

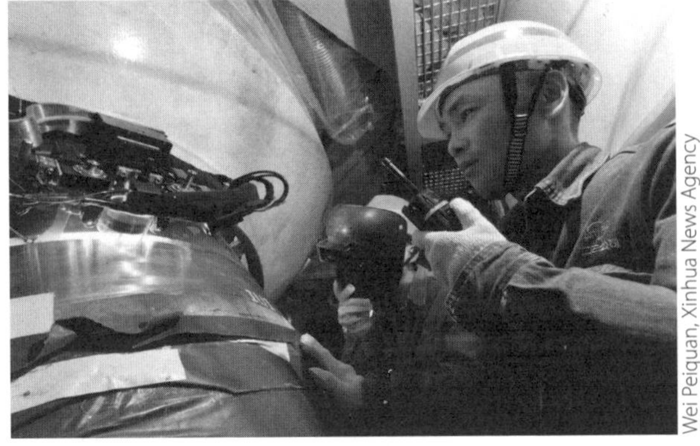

**On 25 January 2011, technicians perform an automatic welding operation at the main channel of the No. 1 generating unit of Ningde Nuclear power station**

Wei Peiquan, Xinhua News Agency

This was a successful implementation of narrow gap automatic welding of primary pipes in a nuclear power plant. It was developed independently by China and was another major step for the country to grasp third-generation nuclear power technology in the future.

During the construction of a nuclear power project, the primary pipes at the core of the reactor are the thick wall pressure pipeline connecting the main equipment such as the reactor pressure vessel, the main pump and the evaporator. It is called "the aorta" of a nuclear plant, shouldering the important safety function of protective screening.

During the construction period, China's nuclear power stations adopted traditional electrode welding for the main pipeline, but the main pipeline automatic welding technology has been applied in other countries.

Since 2008, China Nuclear Power Engineering set up a special team to develop the automatic welding technique. After considerable effort, the team finally finished the development of the technique with an independent IPR. Since 2011, the technology has been applied successfully in the construction of nuclear power stations in Ningde, Hongyanhe, Yangjiang and other places.

### *High temperature gas-cooled reactor*

In everyday language, the reactor is known as an "atomic boiler", a device that generates atomic energy through controlling the reaction of nuclear fuels. Usually, the nuclear fuel of a reactor is Uranium 235, which could lead to nuclear fission if triggered by a neutron. After absorbing a neutron, an atomic nucleus will split into two lighter nuclei, release energy by means of heat and generate two or three new neutrons. Under certain circumstances, the new neutrons could trigger the fission of other Uranium 235 nuclei. If this reaction continues, there is a "chain fission reaction". To achieve chain fission reaction, Uranium 235 must reach a certain quantity, while at the same time the moderator is needed to slow down the high energy neutrons to "hot" neutrons. Nuclear energy can be used if the nuclear fuel's reaction can be controlled to slowly release the nuclear energy and derive the heat out of the reactor heating medium.

Most of the world's reactors use a rod-shaped fuel element. Its heating medium is water, which can reach a maximum temperature of 328°C. The heating medium for a gas-cooled reactor is helium, and graphite is used as a moderator and structural material. A coated spherical fuel element is produced using high-tech equipment and techniques. Its reactor core temperature could reach 1600°C and the temperature of helium outlet is 900°C. This is beyond the capability of any other type of reactors.

**Shidao Bay nuclear plant – first demonstration power plant of a gas-cooled reactor**

Rongcheng Shidao Bay nuclear plant is China's first self-developed demonstration power plant of a gas-cooled reactor. It is also the world's first commercial-scale demonstration power plant of a gas-cooled reactor with a fourth-generation nuclear energy system and security features module. It was jointly constructed and operated by China Huaneng Group, China Nuclear Engineering Group Corporation (CNEC) and Tsinghua University. Phase-I will involve the construction of a 1×20GW Gas Cooled Reactor nuclear power unit and corresponding facilities. Its construction and operation was handled by Huaneng Shandong Shidao Bay Nuclear Power. It cost more than Rmb3bn to build, rising to Rmb5bn when the scientific research input is included.

In July 2011, the 20GW demonstration power plant of the gas-cooled reactor in Rongcheng, Shandong province, passed a nuclear security inspection led by the National Nuclear Security Bureau after the Fukushima nuclear accident in Japan.

On 8 October 2011, research on this demonstration project achieved significant progress. Test production of the spherical fuel element was completed at the Nuclear Institute of Tsinghua University, with the successful manufacturing of an irradiation test sample of the element. The spherical fuel element is the first protective screen of a gas-cooled reactor and a key component that needs continuous production and supply after the demonstration power station is completed and put into operation. The industrialised production capability of a high-quality spherical fuel element is one of the most important tasks for the special major project of a gas-cooled reactor.

The "national special major nuclear power project base", located in an eastern area of Tsinghua University's Institute of Nuclear Technology, has been largely completed. Projects such as the advanced reactor project laboratory, the comprehensive laboratory of the National Nuclear Power R&D Centre and the reconstruction and expansion of a 35kV substation, have been completed, giving the Institute of Nuclear Technology the environment of a leading engineering laboratory. In coming years, a group of world-class engineering experiments will be performed at the base, including the big helium circuit, fuel unloading system, KLAK system, vapour generator, helium-purifying system, and large pressurised water reactor with an experimental table.

So far, more than 90% of the equipment has been ordered for the demonstration project. The pressure vessel in the reactor overcame technical difficulty with a 400-ton forging piece entering the assembly phase. In October 2011, as the last helium gas circulator was put in position, the experimental circuit of the large helium project began to enter the debugging stage. (By Li Yan)

(*Science and Technology Daily* Page 9, 4 March 2012)

## *Knowledge link-I*

### The development process of China's controllable nuclear fusion

- In 1991, China carried out Superconducting Tokamak research, probing how to solve the problem. In 1994, China began to operate the second biggest HT-7 device in the world.
- In 1998, China listed the fully superconducting nuclear fusion experimental device as one of the six major scientific projects during the ninth five-year plan.
- In 2006, Chinese scientists built the world's first nuclear fusion experimental device the Fully Superconducting Tokamak EAST (known as the "artificial sun"), simulating the sun to produce energy.
- From the first round of electric discharge on 28 September 2006 to the second round on 15 January 2007, China's new-generation "artificial sun" EAST (Experimental Advanced Superconducting Tokamak), shows that China stands at the forefront of the world.
- On 21 November 2006, the International Thermonuclear Experimental Reactor (ITER) was officially launched. This remarkable reactor will take 10 years to build. As one of the participating countries, China will shoulder 10% of the responsibility. Scientists estimate that, by 2050, real nuclear fusion could happen.
- On 30 August 2007, the Standing Committee of the 10th National People's Congress held its 29th conference and passed "The Agreement on the Establishment of the International Fusion Energy Organisation for the Joint Implementation of the ITER Project".

## *Knowledge link-II*

### Chronicle of events for the ITER plan

- In 1985, as a symbolic act to mark the end of the Cold War, the Soviet leader Mikhail Gorbachev and the US president Ronald Reagan proposed at the Geneva Summit that the ITER plan should be launched jointly by the US, the Soviet Union, Europe and Japan. In 1990, the ITER was completed.
- In 1998, due to political reasons and a domestic dispute, the US announced its withdrawal from the ITER plan, saying it wanted to strengthen basic research, but Europe, Japan and Russia continued to cooperate and revised the design of the fast reactor, based on new nuclear fission research and other high technologies in 1990s.
- In 2001, Europe, Japan and Russia's joint working group finished a new engineering design (EDA) of ITER and the development of the main components.

- In 2002, based on the EDA, Europe, Japan and Russia began to discuss an international agreement of the ITER plan and the establishment of a corresponding international organisation.
- At the beginning of January 2003, China announced that it had begun participating in the consultation. Then, at the end of January, President Bush announced that the US would rejoin the ITER plan. South Korea was accepted as one side in the consultation of the ITER plan in 2005. In June 2005, the six parties signed an agreement to build the ITER at Cadarache, France's nuclear technology research centre. India joined in the consultation in 2006. Finally, governments of the seven member countries signed an international agreement to develop the ITER on 25 May 2006.

(China Science and Technology Network)

# Chapter 7

# Agricultural science and technology

Technology in China's agricultural sector has developed at a fast speed over the past decade, which has supported the growth of China's rural economy. Remarkable achievements in agricultural technologies have contributed to the steady eradication of poverty in society.

Agricultural technologies have substantially benefited rural production. The successfully launched hybrid rice, hybrid maize, dwarf abortion wheat and double-low rapeseed have enhanced agricultural production capacity.

Technology has enabled special vegetables to be cultivated in a soil-free environment. Greenhouses provide the most favourable growth environment for crops using technologies such as temperature control, irrigation and integration control systems. China's technologies in this area are among the most advanced in the world.

Agricultural technologies raise the possibility of eradicating crop pests, and lowering the loss rate of crops and death of livestock. China recently successfully developed vaccines, including the highly pathogenic avian influenza (HPAI) and Porcine Reproductive and Respiratory Syndrome (PRRS), playing a very important role in controlling diseases. Some technologies, including meteorological satellite remote sensing technology and satellite positioning technology are widely applied, increasing controllability and the possibility of intervention. Recently, China has concentrated on agricultural technology innovation and investment, including the gene transfer project. It has also carried out a plan to build an agricultural technology system and a space-breeding technology system, laying a foundation for improving agricultural technology innovation capabilities and promoting agriculture competiveness.

Agricultural technologies strengthen rural resources and environment technologies and the capabilities of using renewable energy and facilitate agricultural sustainable development. Up to now, production mechanisation of major crops has developed rapidly, in which wheat production is fully mechanised and rice mechanical transplanting, mechanical harvest and maze harvest mechanisation have been enhanced at a fast pace.

## Modern agricultural equipment and cultivation methods

### *Multi-functional agricultural equipment and facilities*

In terms of multi-functional agricultural equipment and facilities, China has developed 105 items of key technologies such as the digital design of agricultural mechanisation, tractor power shift, highly efficient grain harvesting and the bionic drag reduction of machines and tools. It has also developed 118 new products, such as the 200 horsepower tractor and supporting compound machine, which has remarkably enhanced the industry's technological level.

#### Combined tillage machine

Compound minimal tillage land preparation technology is a new type of technology that can be applied to different crop rotation systems and to the dry soil conditions that exists in northern China. This technology combines multi-layer soil structure theory and protective farming techniques, generating a compact upper layer of soil, a loose middle layer and a lower layer that is alternately loose and compact.

#### Large horsepower tractors and compound operation equipment

China has grasped basic technologies such as the digital modelling of agricultural machines, virtual design and reliability enhancement testing (RET), focused on key bottleneck technologies such as load shifting and soil bionic drag reduction, developed a 200hp tractor, supporting compound machines, wide combined soil preparation tools, precision seeding, separated layer fertilisation, intertilling and a wide range of high-efficiency plant protection. They are applied in northeast China and other major grain-producing areas.

#### No-till fertilising seeder

With advanced technology, the 2BMG-20 no-till fertilising seeder can be used in arid and semi-arid regions of China. It is one of China's key technologies for agricultural production in arid areas and provides the necessary equipment for the country's grain-production safety.

### Self-propelled walking brush cutter

The 9GZ-1.0 self-propelled walking brush cutter overcomes the difficulty of cutting desert shrubs, promotes the return of grain plots to forestry and helps control desertification. The equipment incorporates a multi direction auto adjustment copying device, a universal sliding palm and a cutting element that have obtained several national patents. The equipment can undertake the harvesting of desert shrubs, such as caragana microphylla, rose willow and salix mongolica.

### Movable automatic collecting and baling machine

To address the need to process large volumes of crop stalks due to the rapid development of China's biomass energy market, there was a pressing need to develop large, highly efficient stalk-baling equipment for collecting wheat, corn and other crop stalks.

The 9YDF-130 movable automatic collecting and baling machine is mainly used for the large-scale collection of agricultural biomass stalks and pasture, and to pick up, compress and bind the material in a single process.

### Comprehensive cultivation technology and supporting machinery and tools

This series of machinery and tools for tillage and other functions is designed to reduce time and costs, save energy and promote working efficiency. It plays an important role in alleviating soil compaction and reducing water loss and soil erosion. This product won first prize at Heilongjiang Provincial Science and Technology Progress Award in 2010.

## *Modern agricultural practices*

Zhou Mi, Xinhua News Agency

**Technicians direct farmers to use a rice transplanter (19 April 2012)**

**Reducing the need for oxen and manual labour in spring ploughing and rice transplanting**

Traditional ploughing is gradually becoming mechanised and more scientific, resulting in a reduction in the use of oxen and labour during cultivation. The area of mechanical rice transplanting is being enhanced year by year. In addition, seedling slinging and direct seeding have largely taken the place of traditional rice transplanting.

**Plastic film-covered cultivation helps farmers save water and increase efficiency**

## New crop variety breeding

### *New rice variety*

- "Y Liangyou 1" was planted in 2011 in more than 13,000 sq km across the country. In 2012, another 5,000 sq km would be planted. Large fields planted with this new variety can yield 600kg, about 150kg more than China's average rice production level.

- The production of "Y Liangyou 2" per mu (about 667 sq metres) is 926.6kg. This was a world record for hybrid rice.

On 9 September 2011, Hunan Provincial Academy of Agricultural Sciences announced that China had achieved a major breakthrough in the planting of super-hybrid rice. The average production per mu in hundreds of test fields, under the direction of Yuan Longping, exceeded 900kg, creating a world yield record for hybrid rice.

China has always led the world in hybrid rice research. The 900kg per mu yield for large growing areas is a world record that has yet to be broken. It is also a topic for addressing the world's food problem by Chinese experts led by Yuan Longping.

Li Zongxian, Xinhua News Agency

**On 16 April 2012, farmers in Tumen village, Shanting district, Zaozhuang municipality, Shandong province, sowed peanuts using film mulching technology. The farmer is spraying herbicide**

"The goal of 700kg per mu yield took us four years; the goal of 800kg, another four years. For the major breakthrough of 900kg, we took seven years," said Mr Yuan. "The success of 'Y Liangyou 2' proves that, as long as there is a good variety, good cultivation method and high quality farmland, China still has plenty of scope to increase production, space and potential."

The 81-year-old Yuan Longping said he would next set a goal of 1,000kg per mu yield. "We will try to realise this goal in around 10 years."

## Knowledge link

## The story behind "900kg"

On 18 September 2011, in the village of Leifeng, Gu'ao township, Longhui county, Hunan province, after strict procedures of yield measurement and acceptance, an expert group designated by the Ministry of Agriculture showed that the average yield per mu of Y Liangyou 2 in 18 test fields covering 107.9 mu reached 926.6kg. This was a world record in hybrid rice cultivation.

"Mr Yuan once said that hybrid rice comes from the joint efforts of us all. It was, is and will continue to be the case in future," said Deng Qiyun, a student of Yuan Longping and R&D chief of the super-hybrid rice Y Liangyou 2. He was speaking at Hunan Hybrid Rice Research Centre, being interviewed by a Xinhua News Agency journalist on the afternoon on 24 September.

**Yuan Longping checks rice-growing conditions at the side of a test field (May 2007)**

***Condensing generations of painstaking care***

In 1970, Li Bihu, Yuan Longping's assistant, and Feng Keshan, a technician stationed at a farm, discovered a wild abortive rice plant, opening the door to the study of "the three lines" (male sterile line, maintenance line, restoring line).

In 1984, Shi Mingsong, an agricultural technician from Xiantao city, Hubei province, developed a "photoperiod sensitive genie male sterile rice", laying the foundation for research into two-line hybrid rice.

In 1987, under the guidance of Yuan Longping, a young researcher called Deng Huafeng at Anjiang Agricultural School found a "thermo-sensitive genic sterile" rice plant and bred it into the Annong S-1 genic sterile line.

After that, Luo Xiaohe, who is called the 'two-line method founding father", invented the technology for cold water irrigation breeding, resolving the problem of low yield, low temperature, sensitive, sterile breeding.

Under the concerted efforts of coordinating units from across the country, China's research into two-line hybrid rice achieved basic success in 1995.

In 1997, on the basis of the two-line method, Yuan Longping was able to realise the phase I goal of 700kg per mu in 2000 and the phase II goal of 800kg per mu in 2004.

Seven years later, on 18 September 2011, in Leifeng village, when Yuan Longping calculated the final real production, the prospect of realising of 1,000kg per mu in Phase-IV seemed just a matter of time.

"Every achievement in hybrid rice is the result of countless people's efforts," said Deng Qiyun. "The reason why he has come this far is that he stands on the shoulders of giants: without the technical route formulated by Mr Yuan, without the hard work by many predecessors, it's impossible to have the achievements of today."

### *A "not wonderful" true story*

"A wonderful story is not true and a true story is not wonderful." The 49 year-old Deng Qiyun repeated the sentence again and again. In the opinion of Deng Qiyun, an experiment repeated thousands of times really couldn't be described by the word "wonderful".

On 18 September 2011, when the expert group designated by the Ministry of Agriculture undertook field supervision and yield measurement work, Deng Qiyun, who was not there at the time, appeared very confident. "I'm very clear how my child performs."

"The development process of Y Liangyou 2 is my growing process," Deng Qiyun said. "Since starting to study for a doctoral degree under the guidance of Yuan Longping in 1997, I've been following him to carry out research into the breeding of super-high yield hybrid rice."

Deng Qiyun believed that the breakthrough in the form of Y Liangyou 2 proved the veracity of the technological route followed by Yuan Longping.

"He once said the core technique of high-yield hybrid rice is to combine morphological improvement with hybrid vigour." Deng Qiyun quoted an analogy that was drawn by Yuan Longping: a good shape is like the body of basketball star Yao Ming, while the hybrid vigour should be full of physical strength instead of puffiness.

But, it's by no means easy to make a rice seedling grow into an ideotype while also considering population yield.

To breed the new super hybrid rice variety, Deng Qiyun and his colleagues suffered many hardships. "When other people are enjoying air conditioning, we have to observe the rice growing under the scorching sun. We can't even take an umbrella, otherwise we will not see clearly," said Deng. Many researchers in his team could not even return home for a family reunion for several Spring Festivals in a row. As for exposure to strong sunlight and crossing mountains and rivers, that is part of the routine for the researchers.

"Many people don't understand how we could take pleasure in hardship. I think that's because there is always a hope waiting for you there," said Deng.

### *"Questioning could help us make greater progress"*

"Nine hundred kilograms is just a number in the test field, without any practical significance." "The rice quality of the super hybrid rice is not good." "GM elements are probably contained in super hybrid rice." In addition to the national pride generated by the record set by the super hybrid rice, there were also questioning voices.

Responding to such criticisms, Yuan Longping and his team have always kept clear headed.

The test area in Leifeng village, Yanggu'ao township, is more than 300 metres above sea level, with good ecological conditions and sunshine, and a long growth period.

"Here, the breakthrough of 900kg doesn't mean the same results could be achieved on a large scale," said Deng Qiyun. Even following the Phase III goal of super hybrid rice from the Ministry of Agriculture, the target was initially realised only when the same ecological area attained 900kg per mu for two consecutive years.

But this doesn't mean a devaluation of achieving the 900kg goal. It is known that, in 1980, Japan had raised its 15-year plan to increase production by 50% through indica japonica breeding. The International Rice Institute raised a new plant type breeding plan to increase the area yield by tropical japonica rice in 1989. But so far, the goal of the large area production increase hasn't been realised by Japan or the institute.

"This time, the achievement of 900kg per mu in hundreds of mu fields in Longhui county means that China's hybrid rice research has been far ahead of the world," said Deng Qiyun. Even allowing for a 15-25% fall on the trial performance when planting large-scale in less ideal conditions, that still means a level of 700kg per mu can be achieved.

As for questions over the rice quality, Deng Qiyun said the Hunan Hybrid Rice Research Centre accepts visitors from around the world each year. Each time, Yuan Longping asked Deng Qiyun to cook meals using super hybrid rice and ask guests to have a try. Many visitors said there was no significant difference in taste between hybrid rice and Thai rice. As for the possibility of hybrid rice being mixed with GM rice, Deng said that hybrid rice is an improved variety achieved through natural crossing and is a totally different concept from the GM technology of integrating a gene of from one kind of plant into another by means of genetic manipulation.

"The progress we make in hybrid rice needs more understanding and encouragement from society. But we also welcome all kinds of voices," said Deng, "because questioning could encourage us to achieve greater progress."

## *New corn variety*

Head smut is one of the world's most severe and common corn diseases, seriously harming corn production. Scientists make use of a series of molecule marks in a segment of a disease-resistant QTL plant and employ the strategy of backcross breeding to incorporate the disease-resistant trait into elite lines. The disease resistance of the improved inbred lines and their crossbreeds increase at different levels (20-90%), while other positive agronomic traits remain the same.

There are two important high-yield records in terms of corn cultivation: one is a spring corn high-yield record held by the founder of the Pioneer Company, Henry Wallace; the other is the summer corn high yield record held by Li Denham, a Chinese cultivation expert. According to calculations, the corn seeds cultivated by Mr Li raised the capacity of about 67 sq metres of Chinese farmland from feeding one person to feeding 4.5 persons. His corn seeds have been sown widely across China, increasing the direct economic benefit by more than Rmb100bn.

In 2005, the maize variety Denghaichaoshi-1 developed by Li Denghai created a new record of world summer corn at 1,403kg per mu.

In 2007, Li Denghai became the leader of a major project on new super corn variety breeding and industrialisation, which was listed into the 11th five-year plan, the National Sci-tech Support Plan. So far, they have bred more than 10 new super corn varieties, with four passing national or provincial accreditation, including Denghai-662 and Denghai-605 which have passed the accreditation of the National Crop Variety Approval Committee (NCVAC). Denghai-661 has excellent characteristics such as stable yield, ear uniformity, high rate of seeding, high quality, disease- and lodging resistance and a mature living stalk. As a suitable variety for a high yield of summer corn in the Huang-Huai-Hai Plain, China's top corn-growing region, it has significant scope to be grown over a wide area.

### *GM bollworm-resistant cotton*

According to a Xinhua News Agency report on 20 March 2012, China has become a world leader in GM cotton research and has a wholly-owned international invention patent and other independent IPRs, paving the way for it to shake off its reliance on importing high quality cotton. This is another important technological achievement in the high-tech sector.

For a long time, pest and disease damage and low fibre quality have been the major problems affecting the development of China's cotton industry. In the 1990s, Chinese scientists developed the first generation of GM cotton with independent IPR, Bt cotton, which successfully resisted bollworm damage and increased the market share of domestic GM pest-resistant cotton from 5% to more than 95%. Until now, GM pest-resistant cotton has been the GM crop with the largest commercial cultivation area.

Scientists from the Cotton Research Institute under the Chinese Academy of Agricultural Sciences, Peking University, Fudan University, Southwest University and other research units and academies have worked together and combined resources to research China's first generation GM cotton and finally realised a major breakthrough in variety cultivation of high quality GM fibre and material.

The "CRI70" was cultivated by the Cotton Research Institute of the Chinese Academy of Agricultural Sciences (CRI of CAAS) by using molecular aggregation technology to create a hybrid utilising the high fibre qualities of Sea Island Cotton and GM pest-resistant cotton. CRI70 cotton fibre is 32.5mm long, meeting the high-quality cotton standard. The RRM gene high-quality big boll cotton cultivated by CRI of CAAS and Fudan University could have a 7.5 gram single boll, much heavier than for an ordinary cotton variety. The number of bolls borne is 20% higher than for ordinary cotton varieties, giving the potential for high yields and high quality cultivation.

## Cultivation with agricultural technologies

### *Balcony garden: what is the distance between dream and reality?*

There was a time when people were satisfied about growing crops in the on-line game Happy Farm and stealing vegetables from their friends' gardens. As people are worried about fluctuating vegetable prices, overused chemical fertilisers and excessive pesticide residues, more of them want to eat pollution-free, green vegetables.

In major cities such as Beijing, Shanghai and Harbin, "urban peasants" are active in growing green vegetables on their roofs and balconies.

#### Pursuing dreams in vegetable gardens

Wu Ma, who lives in Harbin, shares his planting experience with his "peasant friends" on the "I love gardens" bulletin board system. "My first harvest is lettuce. I make noodles with the lettuce for my breakfast. I was so proud of my first harvest. It was just like my child becoming a champion."

In Beijing, Mrs Wang's dream is to grow her own vegetables, but she had no experience of planting since she has lived in a city since childhood. Now, the 60-year-old Mrs Wu has learned to select seeds, turn soil, trap insects and harvest her own tomatoes, peppers and edible rape. Her dream to become a farmer has been realised without the need to even go outdoors.

Ms Yang is a white collar worker who doesn't have enough time to look after vegetables; she has taken advantage of soilless culture technology. Her vegetables only need sunshine, water and air. When they sprout, they grow in nutrient solutions. This is the easiest way of planting.

The residents of Min'an community in the Dongcheng district of Beijing are growing vegetables together. There are more than 80 rectangular flowerpots on a garden roof that the residents built together. All of them enjoy selecting seeds, fertilising and sharing the harvest.

## The experience of "urban peasants"

What is the cost of a balcony garden? What are the harvests like? The cost of planting varies according to the different growing methods and agricultural technologies. However, every urban peasant has his or her planting experience.

Mrs Wu did the maths: Rmb10 for a flowerpot for 20 lettuce; Rmb8 for a bag of seeds enough for more than 10 flowerpots. In summer, vegetables sprout after a week and become mature after three weeks. Mrs Wu comments that lettuce in markets sell for Rmb2-3. She also says that she spends an hour watering and fertilising, which is a useful form of physical exercise for a retired worker.

Zhang Xu Xinhua News Agency

**A large intelligent multi-span greenhouse was unveiled at the 7th International Strawberry Symposium. Inside the greenhouse, there are internationally advanced agricultural technological equipment, such as a mobile sprinkling irrigation systems, automatic intelligent environmental control systems and ground source heat pumps, which can meet a strawberry's growth environmental needs from seedling to harvest and realise the automatic production of strawberry cultivation**

In comparison with Ms Yang's planting, the soilless culture technology is much more expensive. Her gadgets in the balcony garden consist of white vertical chutes, covering less than one square metre. There is also a container filled with nutrient solutions. Altogether, these items cost Rmb940. The seeds and nutrient solutions cost much less. Ms Yang says that she has planted 50 lettuce and rape plants altogether, which she can cook once a week. For her, sunshine, air and green methods are of vital importance.

According to the experience of urban peasants, the most suitable vegetables in balconies include lettuce, celery, parsley, Chinese cabbage, spinach, tomatoes, cucumber, bitter melon and peppers. The temperature is between 0°C and 25°C. The average balcony is suitable for planting. Vegetable seeds can be bought from local agricultural institutes and seed companies.

**To live green is a life attitude**

The growth in popularity of "balcony gardens" is to be warmly welcomed. However, it will take time to become universally accepted. Shi Yajun, President of Beijing Urban Agriculture Institute, says that in big cities in Japan and Singapore, balcony agriculture is everywhere and has become a part of life for city residents. In China, urban residents are starting to accept pollution-free and environmentally-friendly concepts. It is hard to get everybody moving forwards.

Ju Huanzong, Xinhua News Agency

**In a semi-circular, plastic film-covered greenhouse, rows of vegetables planted in vertical "pillars" grow strongly. These vegetables, using soilless culture, were cultivated by 32-year-old overseas returnee Ye Kaifeng, in Cixi county, Zhejiang province**

Experts believe that the future of balcony gardens is positive. For the average family, normal flowerpots are enough for planting. For the more fashion-conscious, wall-mounted devices can be bought, which are both beautiful and functional. For elderly people, a balcony garden can be a particularly rewarding experience. The balcony garden is never a closet but a platform for people to show their life attitudes.

## Effective water-saving irrigation

Effective water-saving irrigation is the general term for all forms of irrigation except for earth canal watering and surface flooding irrigation. The application of effective water-saving irrigation changes the traditional water conservancy project and improves the earth utilisation ratio. It also changes the traditional way of labouring, helping to reduce costs and enhance efficiency.

### *Sprinkler irrigation*

Compared with flood irrigation, sprinkler irrigation is able to save 30% more water. It is mainly applied in fields and is easily controlled in regions. It also has advantages such as increasing production and the earth utilisation ratio. However, it involves high energy consumption and evaporation. It requires more water and light winds.

### Micro irrigation

Micro irrigation is one of the most advanced water-saving irrigations. It delivers water to the parts of the crop that need it. It turns "watering the earth" to "watering crops". Micro irrigation is applied to facilitate agriculture and economic crops and is suitable for all terrains and soil types. It saves water and increases production. It uses half the amount of water used in sprinkler irrigation. It is easy to integrate water and fertiliser. However, micro irrigation places high demands on water quality and daily systematic maintenance.

**On 27 March 2012, farmers in Liyang town, Xun county, Henan province water wheat with spray irrigation equipment**

Zhu Xiang, Xinhua News Agency

### Drip irrigation

Drip irrigation makes use of plastic pipelines and drippers to channel water to the roots of crops. It is the most effective irrigation in drought areas, where 95% of the water can be effectively used. Compared with sprinkler irrigation, it saves more water and leads to higher production. It can also increase the effectiveness of fertiliser. It is able to be applied to planting fruits, vegetables, economic crops and crops in greenhouses. It is also used in field irrigation in drought areas.

## Animal disease prevention and control

China has taken a world-leading position in research into avian influenza vaccines. Harbin Veterinary Research Institute of the Chinese Academy of Agricultural Sciences was the first in the world to successfully develop an inactivated vaccine for

H5N1 avian influenza and Recombinant Fowlpox Virus Vector Vaccine for the H5 subtype of avian influenza. It was the first to develop the new type vaccine for fighting against both highly pathogenic avian influenza and Newcastle disease.

A breakthrough has been achieved in the prevention and control of Porcine Reproductive and Respiratory Syndrome (PRRS). A research project known as the Systemic Prevention and Control Technology for the PRRS Pathogen Molecular Hereditary Variation Diagnosis is making great progress, through the Institute of Animal Science and the Veterinary Medicine of Shandong Academy of Agricultural Sciences and Qingdao Agricultural University. This achievement has reached an advanced level internationally and was awarded Shandong's Scientific and Technological Progress Prize in 2008.

## Agricultural information in rural areas

During the 11th five-year plan, achievements in agricultural information in rural areas of China have been made in the following four aspects: facilities have been improved; information resources have been developed; rural information services now have extraordinary features; information technologies have been applied to agricultural production.

Information facilities have been constantly improved. For instance, every village is now able to access the internet. By 2010, the percentage of villages with access to the internet was 100%, of which 98% were able to access broadband internet. The number of Chinese rural netizens reached 125m, or 27.3% of the national total. In terms of telecommunication networks, every villager is able to make telephone calls. By 2010, all administrative villages and 94% of villages with more than 20 families had telephones. In terms of the broadcasting network, the

Peng Shaozhi, Xinhua News Agency

**In recent years, Yinchuan in Ningxia Hui autonomous region has vigorously promoted new rural information services. It has accelerated FTTH (fibre-to-the-home) and wireless networking technology, which has helped many rural families better access the internet and realise the rapid development of rural information development. On 26 December 2011, villagers of Erdaogou village, Lingwu city, Ningxia Hui autonomous region surf the web at home**

percentage of villagers who could get access to radio and television rose from 86.02% to 87.68%, respectively, in 1997 to 96.78% and 97.62% in 2010.

Information resources have also been well developed. In terms of information collection, the Ministry of Agriculture has built nearly 40 channels in the agricultural system and set up more than 8,000 stations. Thus, information on production, prices, technology and education, natural disasters, animal disease, income and quality security can be accessed via remote network. Governments have built more than 4,000 agricultural websites covering ministries, as well as provincial, district and county governments. All provincial agricultural departments, three-quarters of district departments and half of the county departments have built their own agricultural information websites. The Ministry of Agriculture has set up a database of agricultural policies, agricultural economy statistics, agricultural technology and education and product prices for 60 industries.

The rural information service has some extraordinary features. At the end of the 11th five-year plan, counties were equipped with information service institutes, towns had information stations and villages had information spots. All provincial agricultural departments had set up sections responsible for information services, while 97% of urban and 80% county agricultural departments had information management and service sections. More than 70% of villages had a total of 1m information service stations, employing 700,000 people. Distinctive agricultural information service modes have also been developed, and innovative activities introduced. For example, Jilin agricultural commission and the local branch of telecoms operator China Unicom successfully created the "12316" hotline for farmers. Zhejiang local governments are spreading agricultural information and providing services regarding products for sale,

**On 18 May 2012, in a high-efficiency agricultural demonstration garden in Boxing county, Binzhou municipality, Shandong province, a staff member debugs monitoring equipment of the agricultural internet of things inside a greenhouse**

Chen Bin, Xinhua News Agency

via the information service platform called "villagers' mailbox", with 2.36m registered users. Shanghai local governments have built platforms to provide villagers with convenient information services, helping them receive information quickly and conveniently. As a less developed province in western China, Ningxia local governments have been assessing the most suitable mode for their agricultural information, working out how to integrate resources, share information and network in the countryside. They were the first to achieve the goal of providing internet access to every county and giving them information service stations. Yunnan local governments have constructed a digital village, completing 1,494 digital village network clusters, covering 16 cities, 129 counties and 1,348 villages, as well as websites for 13,431 administrative villages and 124,206 villages.

Information technology has been applied to agricultural production. For example, meteorological satellite remote sensing technology and satellite positioning technology, geographical positioning technology and automatic control technology have been widely applied by Xinjiang Production and Construction Corps and other state-owned farms in China. The latest technologies have also been used in agricultural areas, such as 3G, the Internet of Things and Cloud Computing. 3G technology has been expolited to popularise agricultural technologies in places as far apart as Daxing, Beijing, Xinghua, Jiangsu, Luohe, Henan and Turpan. The Internet of Things has played an important role in planting, gardening and farming in Heilongjiang, Beijing and Jiangsu. Some dairy farms have introduced advanced milking robots from abroad based on the Internet of Things. These demonstration projects are highly effective and worth promoting.

# Chapter 8

# The earth and the ocean

In recent years, China has made great breakthroughs in deep earth exploration and marine resources development.

The exploration of the earth's crust is a major scientific project designed by Chinese scientists over six years. The SinoProbe – Deep Exploration in China (2008-12) is a research programme of the exploration of the earth's crust. Nine experimental projects of the SinoProbe were completed by the end of 2012.

In terms of offshore oil and gas, in October 2010, China's first 3,000-metre-deep water semi submersible drilling unit "CNOOC 981" was completed. On 9 May 2012, the "Offshore Oil 981" made its first successful drill in the South China Sea and its drill head reached 1,500 metres underwater in the Liwan 6-1 area. It was China's first exploration of offshore oil and gas in deep water, independently marking "a substantial step" for the country's deep sea oil industry.

On 24 June 2012, the Jiaolong, China's manned submersible, plunged to a depth of 7,200 metres in the Mariana Trench. On 27 June, its greatest depth reached was 7,062 meters. It enables China to conduct deep sea scientific research and resources exploration in 99.8% of the world's oceans. This major breakthrough represents China's great progress in deep water technology. As a result, China has the manned submersible, which can dive to the greatest depth in the world, taking new steps in exploring the ocean.

The Chinese Research Vessel "Xuelong" (Snow Dragon) concluded the country's 28th Antarctic expedition and fourth Arctic expedition, setting many new records. The "Dayang Yihao" (Ocean No 1) research vessel ended its 22nd global exploration and returned to its home port in the eastern coastal city of Qingdao on 11 December 2011, completing a 369-day research journey that covered more than 64,162 nautical miles. It was China's largest global exploration and the longest voyage with the largest number of scientists. Its voyage was equal to the times the length of equator. Historic breakthroughs and fruitful results were made during this exploration.

## Deep exploration of the earth

The exploration of the earth's crust is a major scientific project designed and developed by Chinese scientists over six years. The SinoProbe – Deep Exploration in China (2008-12) is a research programme involving an exploration of the earth's crust. The aims of the research work are: to reveal the structure, evolution and dynamic mechanism of the continental lithosphere underneath China; to understand the pulse of crustal activities; to open up ore prospecting in new areas; to understand the physical parameters of the deep crust for national defence; to provide the latest scientific information to achieve a great breakthrough in the search for energy and mineral resources; and to raise the capability for monitoring and warning of geological hazards, so as to accelerate the development of earth science.

Geng Yuhe, Xinhua News Agency

**On the afternoon of 26 July 2011, the Borehole Geophysics Comprehensive Visualiser, due to carry out the long-term observation of the earth's interior, was put into place at the Continuous Station of the Chinese Continental Scientific Drilling Programme (CCSDP), which is located in Maobei village, Anfeng town, Donghai county, Jiangsu province. This set of instruments has a 10-year design life. It can transmit real-time observation data such as borehole seismology and ground temperature to the Institute of Geology, the Chinese Academy of Geological Sciences and the Data Processing Centre of the Earthquake Networks Centre of Jiangsu provincial seismological bureau. It has a great significance for seismological monitoring, research and precise crustal structure in Jiangsu and neighbouring regions**

### *Details about China's deep exploration of the earth*

#### Why conduct a deep exploration project?

The US and the Soviet Union had launched their own deep exploration projects back in the 1960s, with the former discovering oil fields under a mountain range and the latter oil and gas deposits as well as creatures and fresh water underneath the earth. At the start of the 21$^{st}$ century, Australia put forward its

"Glass Earth" plan, which aims to promote understanding of conditions 1,000 meters below the earth's surface.

As energy shortages become more serious and geological disasters occur more frequently, it is becoming increasingly urgent to accelerate the development of earth science. In 2008, the SinoProbe was launched.

"The energy shortage was the first reason for the launch of this programme," said Dong Shuwen, vice president of the Chinese Academy of Geological Sciences (CAGS) in Beijing and director of SinoProbe. The programme includes nine projects, including selected continental scientific drilling and experimenting, experimentation into deep probing technologies and integration, Standard Continental Magneto telluric Network experiment and the China geochemical probe. Altogether, 12 academicians, more than 200 researchers and thousands of scientists have participated in this programme. We hope to have a clear and better understanding of the land underneath China, including the reasons for China's landform, the reasons for earthquakes and the distribution of oil, gas and mineral resources."

SinoProbe, the largest project in China's geological history, is called "China's first step in deep exploration" by the media. All of the nine projects were completed in 2013.

**How to conduct deep exploration?**

Since we cannot see or touch the lithosphere beneath the earth, how do we research it?

The clearest way is deep drilling, according to Yang Jingsui, the person in charge of SinoProbe.

Deep drilling reaches fresh underground material through drilled holes. It can provide the most effective and direct information on the formation of minerals, oil and gas. It also plays an important role in research on earthquakes and volcanic eruptions. For these reasons, the deep drill has been called the super drill.

The super drill, as one of the experimental projects of the SinoProbe, is operating in the Luobusha region of the Himalaya Mountains. Six drill projects have also been operating in Laiyang city, Shandong province, and Tengchong county of Yunnan province. China will select one drill site among all the sites to conduct a 10,000-metre scientific deep drill. Mr Dong said that hundreds of millions of Rmb are needed to dig 10,000 metres under the earth's surface.

Another experimental project has been underway in Ali, Tibet autonomous region, in the area of deep seismic reflection profiling. "Chinese scientists have imaged the lithosphere by exploding dynamite and recording seismic waves from detonations reflected back to the surface," said Chen Xuanhua, director of the general office of SinoProbe at the Chinese Academy of Geological Sciences. "This process is just like having a CT scan for the lithosphere. As long as we have this CT, we can draw a complete map underneath China."

These experimental projects can reveal information about China's ore-concentrated area 3,000 metres underground.

**What achievements have been made?**

SinoProbe was a five-year programme running from 2008 to 2012, and costing Rmb1.1bn.

A journalist asked, since heavy investment has been made in this area, what achievements have been made, and do they have any relevance to people's everyday lives? Yang Jingsui answered by means of an example. "As we all know, ferrochrome is widely used in stainless steel kettles. More than 80% of China's ferrochrome is imported. We discovered the diamond of ferrochrome through the SinoProbe in Tibet's Luobusha area. It provides a scientific basis for discovering large amounts of ferrochrome ore in Tibet."

In terms of mineral exploration, SinoProbe made a 3D image of the metallogenic belt in the middle and lower reaches of the Yangtze River. As a result, an area 3,000-5,000 metres beneath this ore-concentrated area has become visible.

In addition, the earth CT has collected information for about 2,160km of deep seismic reflection profile, which is equal to 48% of the total length of the deep seismic reflection profile (about 4,500km) that China had recorded over the previous 60 years. Based on this information, China can compile a map that will enhance its ability to monitor and predict geo hazards, including earthquakes, as well as providing data and theoretical support.

"These are some of the achievements of SinoProbe," said Mr Dong. "With the development of the SinoProbe, we are making progress in exploring underneath China." He added that the SinoProbe shows that China ranks among the world leaders in terms of deep exploration of the earth.

(Author: Lin Ying and Zhan Yuan, *Guangming Daily*, 15 December 2011)

## Offshore oil and gas exploration

### Oil and gas exploration equipment

A photo of Ocean Oil 981 taken on 9 May 2012

Jin Liangkuai, Xinhua News Agency

### Focus on the first drill of CNOOC 981

On 9 May 2012, the CNOOC 981, China's first self-designed and manufactured deep water semi submersible drilling unit, made its first successful drill in the South China Sea. Its drill head reached the 1,500 metres underwater in the Liwan 6-1 area. It marked "a substantial step" made by the country's deep sea oil industry.

It was the first independent deep water oil drilling to be conducted by a Chinese company and China was the first country to explore for deep water oil and gas resources in the South China Sea. As a milestone, it commenced China's deep water oil and gas resources exploration.

Large deep water drilling rigs are China's mobile national territory and its strategic weapon for promoting the development of the country's offshore oil industry. The drilling of CNOOC 981 opens a new chapter for China's deepwater oil and gas exploration efforts and expands its scope. The drilling will contribute to ensuring the country's energy security and sovereign rights over territorial waters.

The rig is capable of undertaking an offshore operation at a maximum water depth of 3,000 metres and drilling a depth of 10,000 metres. Equipped with third-generation dynamic and global positioning systems, the CNOOC-owned rig represents the highest level of deep water oil drilling in the world. The CNOOC 981 will be mainly engaged in the exploration of drilling, manufacturing drilling, completing drilling and well repair.

It is equipped with a general blow-out prevention system and an automatic closing system that can switch off without electronic and pilot control in case of an accident. Therefore it can efficiently prevent accidents such as the 2010 BP oil spill in the Gulf of Mexico.

Depths greater than 300 metres are internationally recognised as deep waters and those greater than 1,500 metres are ultra deep waters. Globally, 30-40% of marine resources are buried in deep waters and about half of major exploration sites are developed offshore, mainly in deep waters. Deep waters have become an important replenishment area for oil and gas.

The drilling of CNOOC 981 was a milestone in the country's deep water oil and gas exploration efforts. Exploring deep water oil and gas is an effective way to ensure China's domestic oil supply and reduce its reliance on imported oil. We can also improve deep water oil and gas exploration technology and equipment through these efforts, gathering experience for further progress.

The South China Sea is abundant in oil and gas. It is estimated to have 23-30bn tons of oil and 16,000bn cubic metres of natural gas, accounting for one third of China's total oil and gas resources. About 70% of oil and gas reserves in the resource-rich South China Sea are contained in 1.537m sq km of deep water regions. Subject to a lack of key technologies, most of China's current offshore oil exploration is conducted less than 300 metres below the surface. Oil exploration conducted more than 300 metres below the surface is at the initial stage.

In 2006, CNOOC cooperated with Husky Energy to conduct China's first 1,000-metre drill in the Pearl River Mouth Basin and discovered the Liwan 3-1 gas field. This testifies that the deep waters of the South China Sea have the basic geological conditions of a large gas field and are an important replenishment area for oil and gas offshore areas. It also shows that the deep waters of the South China Sea have great potential in the exploration of oil and gas.

In order to make breakthroughs in deep water exploration and development, CNOOC built a deep water fleet based on the CNOOC 981, a 3,000-metre-deep water semi-submersible drilling unit. This fleet includes a deep water pipe-laying crane vessel, an engineering survey ship, a geophysical vessel and a powerful anchor handling tug supply vessel (AHTS). The South China Sea will become the major region for China to explore gas and oil. In 2010, CNOOC's oil output exceeded 50m tons, which is to the maritime equivalent of a Daqing oil field. For CNOOC, deep waters will become an important area for oil exploration in the future and one of the major sources for oil reserves and output increases.

The deep water drilling platform shows that China has taken the lead in Asia in this area. It is of great significance to China to conduct deep sea exploration in the South China Sea. It also lays a solid technology foundation for neighbouring countries to cooperate in the exploration of deep water oil and gas.

## China's submersible Jiaolong

### *China's manned submersible Jiaolong successfully completes programme of deep sea dives*

On 30 June 2012, China's manned submersible Jiaolong successfully completed its programme of deep sea dives, with a sixth and final dive to 7,000 metres.

It also signifies that Jiaolong will now enter a trial period of application after having come through its programme of test dives to depths ranging from 1,000 to 7,000 metres.

The submersible took three plunges to a depth of 7,000 metres during the six dives, the greatest depth being 7,062 metres and each plunge meeting the target. The team has individually tested 289 features of the submersible and 24 functions of the water system, with a particular emphasis on key items.

"The Jiaolong set a national dive record for a manned submersible after reaching 7,000 metres below sea level, which marks a breakthrough and a giant leap in the development of China's deep sea exploration technology. It is a sign that China leads the world in deep sea exploration. It enables China to conduct deep sea scientific research and resources exploration in 99.8% of the world's oceans," said scene commander Liu Feng.

At the same time, the Jiaolong took a number of precious samples from the sea floor, including creatures, sediment and water, and it also took a video. This is the first precious material that scientists have taken from 7,000 metres below sea level in the Mariana Trench, which shows that the Jiaolong has great ability to work on the sea floor.

The test found great biological and geological diversity, showing the mysteries of the deep sea that the international scientific community had not yet discovered. Liu Feng said," The successful test of the Jiaolong provides important technology for scientists to reveal the mysteries of the deep sea. It also shows that the Jiaolong has great application potential."

According to reports, during the trial dives, all eight Chinese self-trained pilots dived to 7,000 metres below sea level. So far, China has eight pilots among the world's 11 pilots who operate manned submersibles at this depth.

By 2012, the Jiaolong had been developed for 10 years and had gone through four years of trial dives from 2009. The depths to which the craft descended have been gradually increased over time, from 50 metres to 1,109 metres, 3,759 metres, 5,188 metres and finally 7,000 metres. During this process, the Jiaolong team has overcome many difficulties in manpower, material resources and finance, completing a major task that no one has ever done before.

### *Localisation rate of Jiaolong is 58.6%*

The Jiaolong is a deep sea manned submersible designed by, and made in China. The Jiaolong was designed independently by Chinese engineers, from scheme design to preliminary design and detailed design. Its key core technologies, such as pressure structure, life support, long range underwater communication and system control were designed and made by Chinese engineers. Its assembly and trial tests were also performed by China independently.

According to the introduction, all the Jiaolong's key components and equipment will be made in China in the future. In terms of the proportion of components, 58.6% of the Jiaolong is home made. This means that China's manned submersible will no longer be controlled by others. The Jiaolong is a true Chinese dragon!

### *Foreign deep submersibles do not dive below 6,500 metres*

China is the fifth country to successfully develop a deep sea manned submersible. Jiaolong belongs to the first tier in the global manned submersible. Currently, there are 90 manned submersibles in use in the world, of which only 12 can reach a depth beyond 1,000 metres and even fewer are capable of going deeper. Countries owning manned submersibles that can operate at a depth of 6,000 or more include China, the US, Japan, France and Russia. The maximum working depth of the manned submersibles of the other four countries, excluding China, is not beyond 6,500 metres and their usual diving depth is no more than 5,000 metres.

## Knowledge link

## The unknown deep diving secrets of Jiaolong

On 24 June 2012, the manned deepwater submersible Jiaolong successfully went below a depth of 7,000 metres, the lowest submerged depth achieved by any submersible of this type in the world. Viewing photos sent back from the site, the Jiaolong was lowered into the water and suspended from cables tied to the mother vessel. On such deep dives, is it necessary to have a 7,000-metre cable? Is the mother ship able to watch the oceanauts' every action through video? Will the oceanauts experience weightlessness just like astronauts?

Ji Guo, the director of the Oceanauts Management Office of the North China Sea branch of the State Oceanic Administration explained the secrets about the deep diving by the Jiaolong.

Luo Sha, Xinhua News Agency

**On 24 June 2012, the Jiaolong rises triumphantly from the 7,020-metre seabed**

### *Jiaolong's unpowered independent submerging and raising*

Journalist: Seen from an onsite photo, the Jiaolong is lowered into the water and suspended from cables of the mother vessel. Will the cable always be attached while the Jiaolong is in its submerging process?

Ji Guo: The Jiaolong performs unpowered, independent submerging and rising. When it is in the water, the frogman takes a rubber dinghy to unfasten the cable. Then the Jiaolong has a completely independent and free operation.

Before every diving experiment, the field men measure the seawater density of the seabed operation area to determine how much kentledge (weights used as ballast in a vessel) should be carried by the Jiaolong. Due to the kentledge, the submersible begins to sink on entering the water. When it reaches a certain depth, some kentledge should be removed so as to make the density of the submersible similar to that of the seawater and reduce the working pressure on the propeller.

When the Jiaolong touched the sea floor, the oceanauts manipulate the submersible to deploy a marker and carry out sediment sampling and a micro-topographical survey. When all the work is finished, the oceanauts remove some more kentledge, making the submersible rise to the surface.

Kentledge is attached to the two sides of the submersible. It is installed only before each submerging experiment.

### *Jiaolong unable to send back real-time video or image*

Journalist: videos and images of astronauts aboard Shenzhou-IX can be sent back in real time. Can the Jiaolong also send back videos and images in real time?

Ji Guo: During the submerging, the Jiaolong relies on underwater acoustic communication apparatus to transmit information. But the properties of underwater sound reduce audio speed and volume, making it impossible to supply real-time audio, text, data or video transmissions.

Sound is transmitted in water at 1,500 metres per second, but as depth increases, the transmission speed will decrease. As the depth increases, the sound's transmission speed will see an inflection point. Therefore, before the submerging experiment, oceanauts should calculate the depth of this inflection point based on seawater salinity data and other factors. The underwater acoustic communication apparatus should be placed below the inflection point depth to ensure decent communication quality.

Through underwater acoustic communication apparatus, the mother vessel and the submersible can establish voice communication. All kinds of information about the submersible can be transmitted back to the mother vessel, such as density, battery volume, oxygen levels inside the module and temperature. But due to the slow speed of sound transmission, a time difference will appear.

### *Oceanauts don't need to wear space suits*

Journalist: When oceanauts emerge from the module, we see that their uniform looks quite similar to that of astronauts. Should they also wear a space suit? Do they experience weight loss inside the module?

Ji Guo: For every 10 metres under water, the pressure on the Jiaolong will rise by one atmosphere. But inside the module, there is an almost constant temperature, pressure level and supply of oxygen. Therefore, oceanauts don't need to wear space suits and they would not experience weightlessness during the diving process.

During the preparatory phase after entering the water, the sea surface temperature is relatively high, as it is inside the module. During the submerging process, the environmental temperature will go down gradually and the temperature in the module will also begin to fall. Therefore, oceanauts need to have installed temperature-measuring instruments during the submerging and wear a thermal diving uniform.

The 7,000-metre sea experiment takes more than 10 hours but the oceanauts are highly focused and it's inconvenient to go to the toilet. Therefore, they consume very little, only eating some apples or chocolate.

# The Snow Dragon scientific expedition

## *Fourth Chinese Arctic research expedition creates new records*

China's fourth Arctic research expedition achieved good results, creating several new records.

During this expedition, China satisfied its desire to reach the North Pole and carry out its own scientific investigations, thus achieving a historic breakthrough. For the first time, China placed an ice buoy on the ice surface of the North Pole and launched a disposable CTD profiler, used to determine the conductivity,

**China's fourth Arctic research expedition team arrive at NL 88°22′ on the Snow Dragon, the highest latitude that Chinese vessels have so far sailed**

Yao Fan, Xinhua News Agency

temperature and depth of the ocean. It also performed ecological observations, collected a large quantity of sea ice and sea water samples, and acquired a 2.5-metre-long North Pole ice core sample for the first time. The expedition was also the first to complete 24 hours of continuous oceanographic observation at a depth of 3,742 metres in the Bering Basin, extended China's marine survey to a high latitude deepwater plain in the Arctic Ocean and gained precious data based on atmospheric physics and chemistry observations. The team were the first to use a multi-plankton sampler to take precise samples at a depth of 3,000 near the North Pole at a latitude of 88°26′.

China's fourth Arctic research expedition covered an area that is 2,300 nautical miles from north to south and 1,100 nautical miles from east to west, involving regions such as the Bering Sea, Bering Strait, Chukchi Sea, Canada Basin, Mendeleev Ridge, Fletcher deep sea plain area and the North Pole. The investigation area, along with the quantity of data and samples acquired, were larger than all the previous Chinese Arctic investigations.

The team completed oceanographic surveying at 135 sites, comprehensive observation of sea ice air at one long-term ice station, eight short-term ice stations and one North Pole site. The quantity and scope of observation sites exceeded the original plan. More samples were collected in this marine ecosystem investigation than all previous Arctic scientific investigations. Initial statistics show that the team collected 1,077 samples of marine phytoplankton, marine zooplankton, microorganisms, meiobenthos, macrobenthos and nekton, as well as more than 2,400 samples of chlorophyll and phytoplankton.

During the investigation process, the team also retrieved an ocean observation submerged buoy system placed there by the third Chinese Arctic scientific investigation team. This was the first set of deepwater submerged buoys longer than 1,300 metres placed by China in the Arctic region. It was also China's first long-term Arctic submerged buoy to involve an observation period of more than one year. The whole retrieving process took only one hour and 50 minutes, a record for China, and the retrieved equipment was of the highest value ever recorded.

During the investigation, the Snow Dragon sailed to its northernmost point of 88°22′ and drifted with the ice to 88°26′, creating a new record in China's seafaring history.

## *Antarctic expedition team finishes multi-disciplinary comprehensive oceanic investigation*

After 13 days of hard work, the 28th Antarctic expedition team finished the multi-disciplinary comprehensive oceanic investigation of seas around the Antarctic peninsula on the evening of 29 January 2012 local time (the morning of 30 January 2012, Beijing time), with 46 working sites. This was the first time that China had undertaken such a large scale oceanic investigation in the seas around the Antarctic peninsula.

A multi-disciplinary comprehensive oceanic investigation involves physical oceanography, marine geology, marine geophysics, marine chemistry, marine biology and ecology. It is aimed at systematically mastering basic environmental information and first-hand data of the investigation area, such as marine hydrology, marine meteorology, marine acoustics, sea ice, marine deposits, pale environment and pale climate records, geophysical field and geological structure, marine chemistry and biology. The team began their intensive operations on 16 January 2012.

"The multi disciplinary comprehensive oceanic investigation fills the gaps for China on the in-depth and comprehensive observation of a cross-section of the Antarctic Ocean. It is the first comprehensive observation with clear goals, strong continuity, wide coverage and highlights, under the guidance of long-term scientific planning," said Jiao Yutian, leader of the team.

### Fruitful results of comprehensive investigation

As a piloting voyage since the implementation of the major special project of north and south poles environmental investigation during China's 12th five-year plan, it has realised several "firsts".

On the morning of 17 January 2012, Beijing time, the team performed deepwater trawling of the zooplankton layers. This was China's first trawling in the Antarctic Ocean, and the five subs deployed were able to collect different zooplankton samples from the sea surface to the seabed. These samples can be used to study the vertical distribution of zooplankton. On 18 January, Beijing time, the team used marine magnetometry in the seas around the Antarctic peninsula. This was the first time in 20 years that China carried out such measuring in Antarctica. Its purpose was to probe the types of seabed rocks and their distribution, to provide a basis for studying the geological structure and its evolution of the Antarctic continent and the surrounding seas. This magnetic measurement involved eight survey lines, with a total length of finished survey lines of about 1,045km.

During the investigation process, the team were the first to use an advanced high-precision sensor for measuring temperature and salinity to ensure the reliability of data quality. Jiao Yutian said, "Its advantages are high precision, good stability, good comparability and the control of data quality."

Captain Shen Quan said that, because it was the first large-scale oceanic investigation in the seas of the Antarctic peninsula, the Snow Dragon research vessel chose sailing routes and the latest weather reports, so as to minimise the influence of adverse weather conditions. The experiences accumulated on sea conditions and weather during the investigation process provided a reference for compiling a special survey plan in the future.

**Keep working in rough seas**

In January, cyclonic activity in the Antarctic Peninsula seas is quite common. Every two days or so, there was a cyclone affecting the sea area where they worked. The average wind force was six to seven on the Beaufort scale, with average wave heights of 2.5 to 3 metres. However, so long as conditions permitted, the 24 team members would lose no time working despite the strong winds and large waves.

Tight schedules and demanding tasks were a feature of this oceanic investigation. Due to the involvement of several disciplines, the investigations were done both on the mid deck and the afterdeck. The mid deck was mainly used to conduct physical oceanographic surveys such as sea water temperature, salinity and flow velocity measurements, water sampling and also to collect zooplankton and phytoplankton samples through biological vertical trawling. The afterdeck was used mainly for carrying out the deepwater trawling of zooplankton layers, benthic trawling, krill trawling, seabed mud and seabed suspended water samples. When arriving at a working site, no matter if it was day or night, the team members were busy on deck.

To ensure completion on time, the team did not take a rest even during the Spring Festival holiday period. Before dawn on the Lunar New Year's Day, the team started the first operation in the Year of the Dragon under an enveloping snow. After a long period of hard work, some team members became so exhausted that they slept in their work clothes in the laboratories. Despite their fatigue, they were content in the knowledge that they had acquired a satisfactory sample.

**Rich samples for reading the sea**

The investigation area is located in the north of the Antarctic peninsula and the northwestern part of the continental shelf and slope, between longitude 44°W and 66°W, with the outflowing area of the bottom water of the Weddell Sea to the east of the South Orkney Islands as the eastern boundary.

Through intensive professional observation, the team completed the water collecting operation in 42 sites on physical oceanography. In terms of marine chemistry, the team acquired more than 3,000 samples in total, of which 406 were analysed. In terms of marine biology, the team collected 88 zooplankton samples while sailing at high speed in a certain sea area, 17 krill samples while sailing at high speed around the Antarctic continent, performed parameter measurements of chlorophyll, temperature and salinity in synchronisation with sampling, and implemented vertical trawling of plankton at 40 sites in the surveyed areas of the Antarctic peninsula. In terms of marine geology, the team collected five seabed sediment cores, of which the longest was 3.2 metres. These data and samples provide clues to the study of the oceanic environment around the Antarctic continent, and give an insight into the interaction of sea, ice and air in the Antarctic region and its influence on global climate, and China's climate in particular.

Taking the seabed sediment core as an example, team member Han Xibin said the sediment core could record information on the evolution of climate and environment, just like the sediments of inland lakes, stalactites in karst caves and tree rings. The 3.2-metre-long seabed sediment core could help study the sea region's history going back at least 300,000 years on paleoclimate, paleoenvironment and paleo ocean, to understand climate events, changes in paleocurrents, sediment fluxes and tectonic activities of the Antarctic peninsula.

Jiao Yutian said, "The multi disciplinary comprehensive investigation in the Antarctic peninsula seas not only helps to build up a stable polar marine scientific research team, but also can enrich China's existing Antarctic observation materials, so as to improve the R&D ability of key technologies of polar marine observation and establish China's long-term observation network of the seas in Antarctica."

## Dayang Yihao (Ocean No.1) round-the-world scientific probe

### *China completes its largest round-the-world ocean probe*

On 11 December 2011, a Xinhua News Agency report stated that China's research vessel, "Dayang Yihao" (Ocean One) returned to Qingdao on that day, completing China's $22^{nd}$ ocean probe. This ocean probe focused on subjects ranging from the deep sea environment, sulphide in the seabed, hydrothermal activity in the Pacific, bio genes and biodiversity. It was a highly comprehensive ocean survey voyage, which covered many areas, a wide span of space, a long period of time and many applications of advanced equipment and reserves. Rich results were achieved in every section of the voyage.

#### Largest scale

This ocean probe involved, for China, its longest voyage, widest probe area and greatest number of participants. Dayang Yihao set off on 8 December 2010 from Guangzhou. It went through nine voyage segments, 369 days and 64,162 nautical miles, equivalent to three times around the equator. The surveyed area covered the Indian Ocean, the Atlantic Ocean and the Pacific Ocean, and 218 researchers from 32 R&D institutions and academies across the country joined in the marine investigation.

#### Great significance of probe results

Researchers found 16 seafloor hydrothermal sites in total during the voyage, accounting for nearly half of the sites known by China (so far, it has discovered 33 seafloor hydrothermal sites in the Indian Ocean, the Atlantic Ocean and the Pacific Ocean).

The hydrothermal sulphide at the seafloor hydrothermal sites is a kind of seabed mineral deposit causing increasing international concern. It is caused by the underwater infiltration of seawater through crustal rifts. Then it is heated by lava, melts the metals in the surrounding rock stratum, such as gold, silver, copper, zinc and lead, and then erupts from underground. These metals form sulphide after a chemical reaction and are then deposited on the seabed. Because they pile up in columns, they are figuratively known as black chimneys. The hydrothermal deposit contains many kinds of metallic ores with economic value and therefore has great significance for a country's development.

During this scientific probe, researchers observed hydrothermal organisms such as prawns, sea crabs, sea anemones and tube worms. They also caught deepwater fish and a large quantity of hydrothermal blind shrimps that were suspected of being a new species identified for the first time in the South Atlantic Ocean. The discoveries were of great significance for the ecological study of hydrothermal areas.

Due to the high temperature, high pressure, highly toxic and dark environment on the seabed, the unique genes of seabed organisms could be used for treating environmental pollution and in other areas. Some experts pointed out that the special environment of the seabed hydrothermal area is similar to that of the early earth. These microorganisms could be similar to the life forms of early earth and therefore might provide a reference for studying the origin of life.

## Self-developed, high-tech equipment plays a key role

The high-tech equipment developed by China played a key role in the investigation. The deepwater comprehensive moored buoy observation system, the pull type resource detection system and the medium-length hole core sampling drill were deployed on a voyage for the first time and with success. Major high-tech equipment, such as the remotely operated vehicle (ROV) and the deepwater acoustic deep towing device, were successfully applied in the hydrothermal area.

## Knowledge link

## Dayang Yihao research vessel

The research vessel Dayang Yihao, or Ocean One, is a 5,600-ton ocean scientific vessel, and China's first modern comprehensive ocean research vessel. In July 1994, it was bought from Russia's national marine geology investigation bureau by COMRA (China Ocean Mineral Resources Research and Development Association) and was renamed as Dayang Yihao after preliminary modification. From 1995, it began to carry out the research and survey of China's ocean mineral resources. In December 2002, after one year was spent carrying out modifications and modernisation, it became an internationally advanced scientific research vessel. From April 2005 to January 2006, Dayang Yihao completed China's first round-the-world-ocean investigation. It was capable of meeting China's R&D needs in international seabed resources for the coming 10 to 15 years.